The Physiology of th

Annotated diagrams of the mechanics of the joints

I. A. KAPANDJI
Ancien Chef de Clinique Chirugicale
Assistant des Hôpitaux de Paris
Membre Associé de la Société Française D'Orthopédie et de Traumatologie

Translated by
L. H. HONORÉ, B.Sc., M.B., Ch.B., F.R.C.P.(C)

Preface by
The late PROFESSOR G. CORDIER
(formerly Dean of the Faculty of Medicine of Paris)

Second Edition Reprint

Volume 2
LOWER LIMB

1 The Hip
2 The Knee
3 The Ankle
4 The Foot
5 The Plantar Vault

With 618 illustrations by the Author

CHURCHILL LIVINGSTONE
EDINBURGH LONDON AND NEW YORK
1970

To My Wife

PREFACE TO THE FRENCH EDITION

This work belongs to a series of three volumes of which the first, on the upper limb, has had well-deserved success.

The same original approach has been adopted in this volume, devoted to the lower limb. The functional anatomy is clearly and precisely set forth with the help of six hundred and eighteen diagrams. The explanatory notes on the mechanics of the joints and the physiology of muscle action are at once brief and perfectly clear.

This new method makes the study of the anatomy and physiology of joints logical and simple. It will appeal to a wide public ranging from the medical student to the physiotherapist and the orthopaedic surgeon.

DOYEN GASTON CORDIER

CONTENTS

THE HIP 9

Movements of the Hip and their Ranges 10
Articular Surfaces and Structure of the Hip 24
The Capsule and Ligaments of the Hip 32
Coaptation of the Articular Surfaces 44
Flexor and Extensor Muscles 48
Abductor Muscles and the Transverse Stability of the Pelvis 52
Adductor Muscles 58
Rotator Muscles 62
Inversion of Muscular Action 66

THE KNEE 72

The Axes of the Knee 74
Movements of the Knee and their Ranges 76
The General Structure of the Lower Limb 80
Articular Surfaces 82
Movements of the Articular Surfaces during Flexion and Extension 88
Movements of the Articular Surfaces during Axial Rotation 90
The Capsule and the Infrapatellar Fold 92
The Menisci and their Function 96
Movements of the Patella on the Femur and Tibia 102
The Collateral Ligaments: their Function and the Transverse Stability of the Knee 106
Anteroposterior Stability of the Knee 112
Cruciate Ligaments and their Function 114
Rotational Stability of the Knee during Extension 124
Extensor Muscles of the Knee 126
Flexor Muscles of the Knee 130
Rotator Muscles of the Knee 132
Automatic Rotation of the Knee 134

THE ANKLE 136

Movements of the Ankle and their Ranges 136
Articular Surfaces of the Ankle 142
Ligaments of the Ankle 144
Anteroposterior and Transverse Stability of the Ankle 146
Tibiofibular Joints and their Function 150

THE FOOT 154

Movements of Longitudinal Rotation; Side-to-side Movements 156
Subtalar (Talocalcanean) Joint 158
Transverse Tarsal (Midtarsal) Joint 162
Movements of the Subtalar and Transverse Tarsal Joints 166
Anterior Tarsal and Tarsometatarsal Joints 174
Extension of the Toes 176
Interosseous and Lumbrical Muscles 178
Sole of the Foot: Muscles and Fibrous Tunnels 180
Flexor Muscles of the Ankle 184
Extensor Muscles of the Ankle 186
Abductor-Pronator Muscles 192
Adductor-Supinator Muscles 194

THE PLANTAR VAULT (Arches of the Foot) 196

General Architecture of the Plantar Vault 198
The Three Arches of the Plantar Vault 200
Distribution of Stresses and Static Distortions of the Vault 206
Dynamic Changes of the Vault during Walking 208
Dynamic Changes Related to the Medial and Lateral Inclination of the Leg on the Foot 210
Adaptation of the Plantar Vault to the Ground 212
Claw Feet (Pes Cavus) 214
Flat Feet (Pes Planus) 216
Imbalance of the Anterior Arch 218

References will follow at the end of the last volume

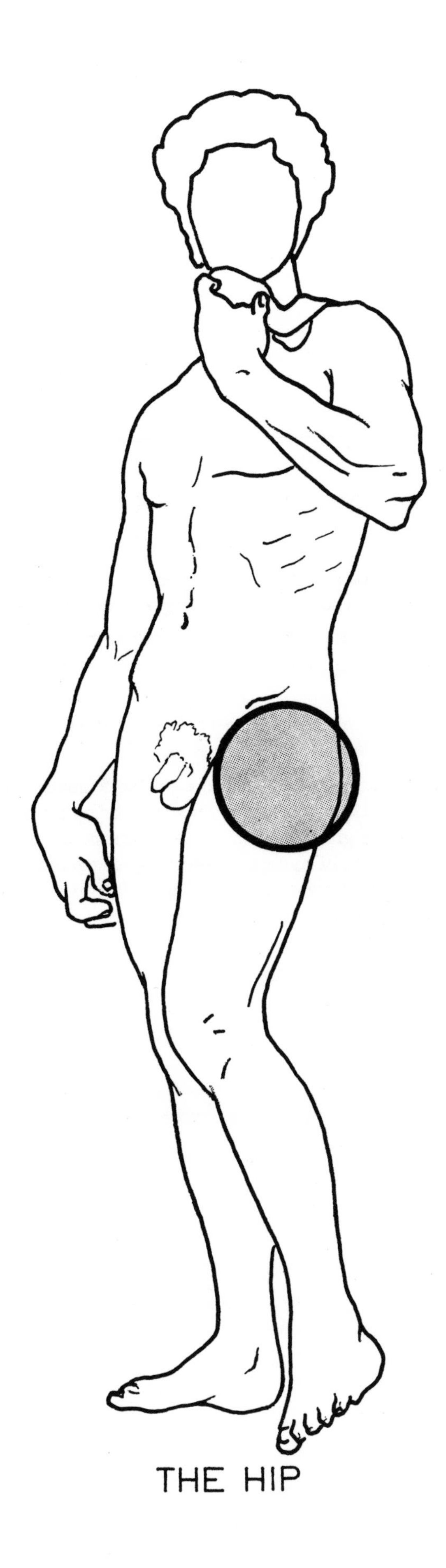

THE HIP

THE HIP

MOVEMENTS OF THE HIP AND THEIR RANGES

The hip is the *proximal joint* of the lower limb and, being located at its root, it allows the limb to *assume any position in space*. Hence it has **three axes** and **three degrees of freedom** (fig. 1).

A **transverse** axis XOX′, lying in a frontal plane and controlling movements of *flexion and extension*.

An **anteroposterior** axis YOY′, lying in a sagittal plane and controlling movements of *adduction and abduction.*

A **vertical** axis OZ, which coincides with the *long axis of the limb* OR when the hip joint is in the 'straight' position. It controls movements of *medial* and *lateral rotation.*

The movements of the hip occur at a single joint: the *hip joint* (coxo-femoral joint). It is a **ball-and-socket joint** with a marked degree of interlocking and in this respect it differs from the shoulder joint which is an open ball-and-socket joint showing great freedom of movement at the expense of stability. The hip joint therefore has a more limited range of movement—partially compensated for by movements of the lumbar vertebral column—but is distinctly *more stable*, being in fact the most difficult joint to dislocate. These features of the hip joint derive from the two basic functions of the lower limb: *support of the body weight* and *locomotion*.

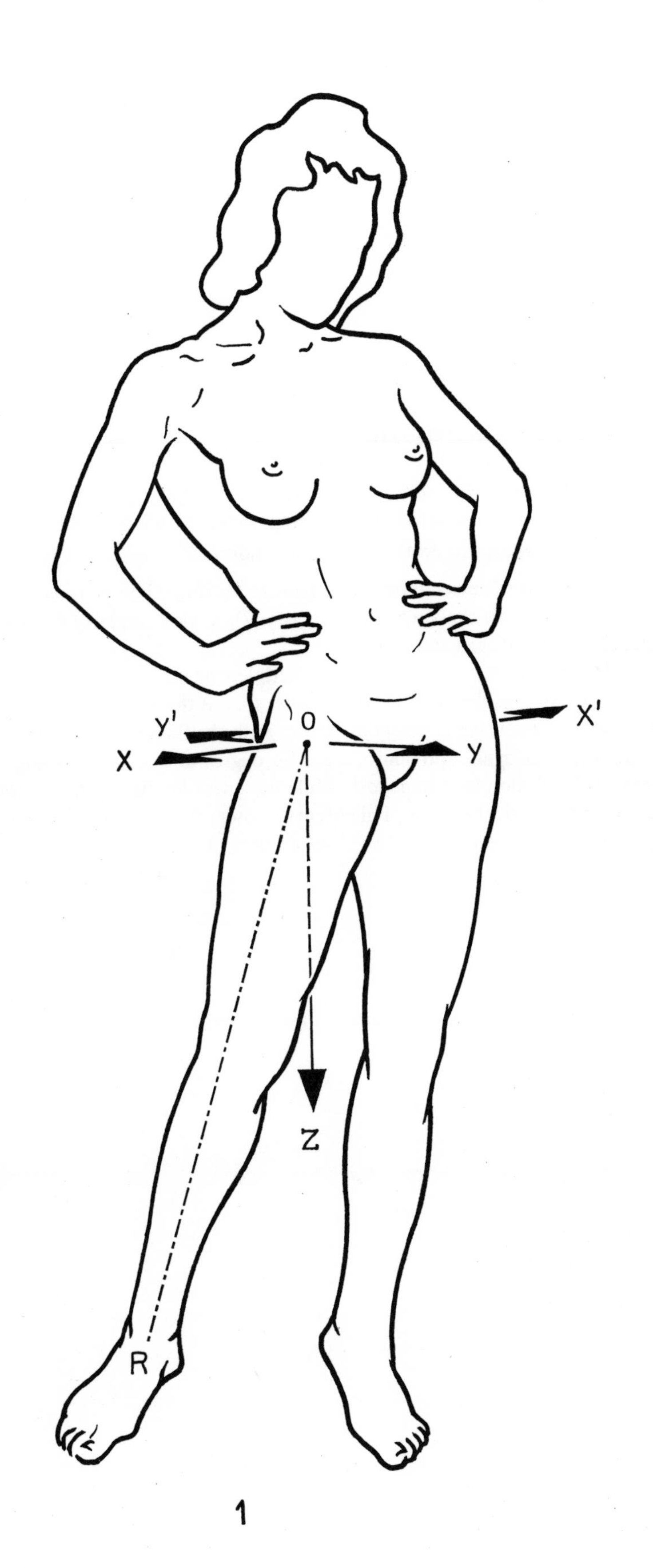

1

MOVEMENTS OF FLEXION OF THE HIP

Flexion of the hip joint is *the movement which approximates the anterior aspect of the **thigh to the trunk*** so that the whole lower limb comes to lie anterior to the frontal plane, which traverses the joint.

The **range of flexion** varies according to the following conditions:

On the whole, *active* flexion is of lesser range than passive flexion. The *position of the knee joint* also determines the range of flexion: with the knee extended (fig. 2), flexion reaches 90°; with the knee flexed (fig. 3), flexion can reach up to 120° or even beyond.

The range of *passive* flexion always exceeds 120° but is still dependent on the position of the knee. If the knee is extended (fig. 4), the range of flexion is clearly smaller than if the knee is flexed (fig. 5), in the latter case the range exceeds 140° and the thigh is nearly in contact with the thorax. It will be shown later (p. 130) how knee flexion relaxes the hamstrings and allows a greater degree of flexion at the hip.

If both hips undergo passive flexion simultaneously while the knees are flexed (fig. 6), the anterior aspects of the thighs come into contact with the chest. This occurs because flexion of the hip is compounded with posterior tilting of the pelvis due to *flattening of the lumbar curve* (arrowed).

90º
2
120º
3
6
145º
5
4

MOVEMENTS OF EXTENSION OF THE HIP

Extension takes the lower limb posterior to the frontal plane.

* The range of extension is notably less than that of flexion and is limited by the tension of the *iliofemoral ligament* (p. 36).

Active extension is of lesser range than passive extension. When the knee is in extension (fig. 7), extension of the hip has a greater range (20°) than when the knee is flexed (fig. 8): this follows from the fact that the hamstrings lose some of their efficiency as extensors of the hip because their contraction has largely been utilised in flexing the knee (p. 130).

Passive extension attains a range of 20° when one bends forwards (fig. 9): it reaches 30° when the lower limb is forcibly pulled back (fig. 10).

Note that extension of the hip is appreciably increased by anterior tilting of the pelvis due to *exaggeration of the lumbar lordosis.* This contribution of the lumbar vertebral column to this movement of extension can be measured (figs. 7 and 8) as the angle between the vertical (fine broken line) and the 'straight' position of the hip (heavy broken line). This 'straight' position is easily determined because the angle between that position of the thigh and the line joining the centre of the hip and the anterosuperior iliac spine is a constant. However, this angle varies with the individual as it *depends upon the orientation of the pelvis,* i.e. the degree of anteroposterior tilting.

The values of the various ranges given apply to the 'normal' untrained subject. They are considerably increased by exercise and training. Ballerinas, for example, commonly do the splits in an anteroposterior direction (fig. 11), even, without resting on the ground; this is due to enhanced flexibility of the iliofemoral ligament. However, it is worth noting that they compensate for the inadequate extension of the posterior limb by an appreciable degree of anterior tilting of the pelvis.

20º
10º
7
8
20º
30º
9
10
11

MOVEMENTS OF ABDUCTION OF THE HIP

Abduction is the movement of the lower limb **directly laterally** and away from the plane of symmetry of the body.

It is theoretically possible to abduct only one hip but *in practice abduction at one joint is automatically followed by a similar degree of abduction at the other joint.* This becomes obvious after 30° abduction (fig. 12), when one first clearly notices tilting of the pelvis, as judged from the displacement of the line joining the surface markings of the two posterior iliac spines. If the long axes of the lower limbs are produced they intersect on the line of symmetry of the pelvis. This indicates that in this position each limb has been abducted 15°.

When **abduction reaches a maximum** (fig. 13), the angle between the two lower limbs is a right angle. Once more abduction can be seen to have occurred symmetrically at both joints so that each limb has a maximum of 45° abduction. The pelvis is now tilted at an angle of 45° to the horizontal and 'looks' towards the supporting limb. The vertebral column as a whole makes up for this pelvic tilt by bending laterally towards the supporting side. Here too the vertebral column is seen to *be involved in movements of the hip*.

Abduction is checked by the impact of the femoral neck on the acetabular rim (p. 32), but before this occurs it has usually been restrained by the adductor muscles and the ilio- and pubo-femoral ligaments (p. 40).

Training can notably augment the maximal range of abduction, e.g. ballerinas who can achieve 120° (fig. 14) to 130° (fig. 15) of *active* abduction without any support. For *passive* abduction trained subjects can attain 180° abduction *by doing the splits sideways* (fig. 16, a). In fact, this is no longer pure abduction since, to slacken the iliofemoral ligaments, the pelvis is tilted anteriorly (fig. 16, b) while the lumbar vertebral column is hyperextended i.e. the hip is now in a position of abduction and flexion.

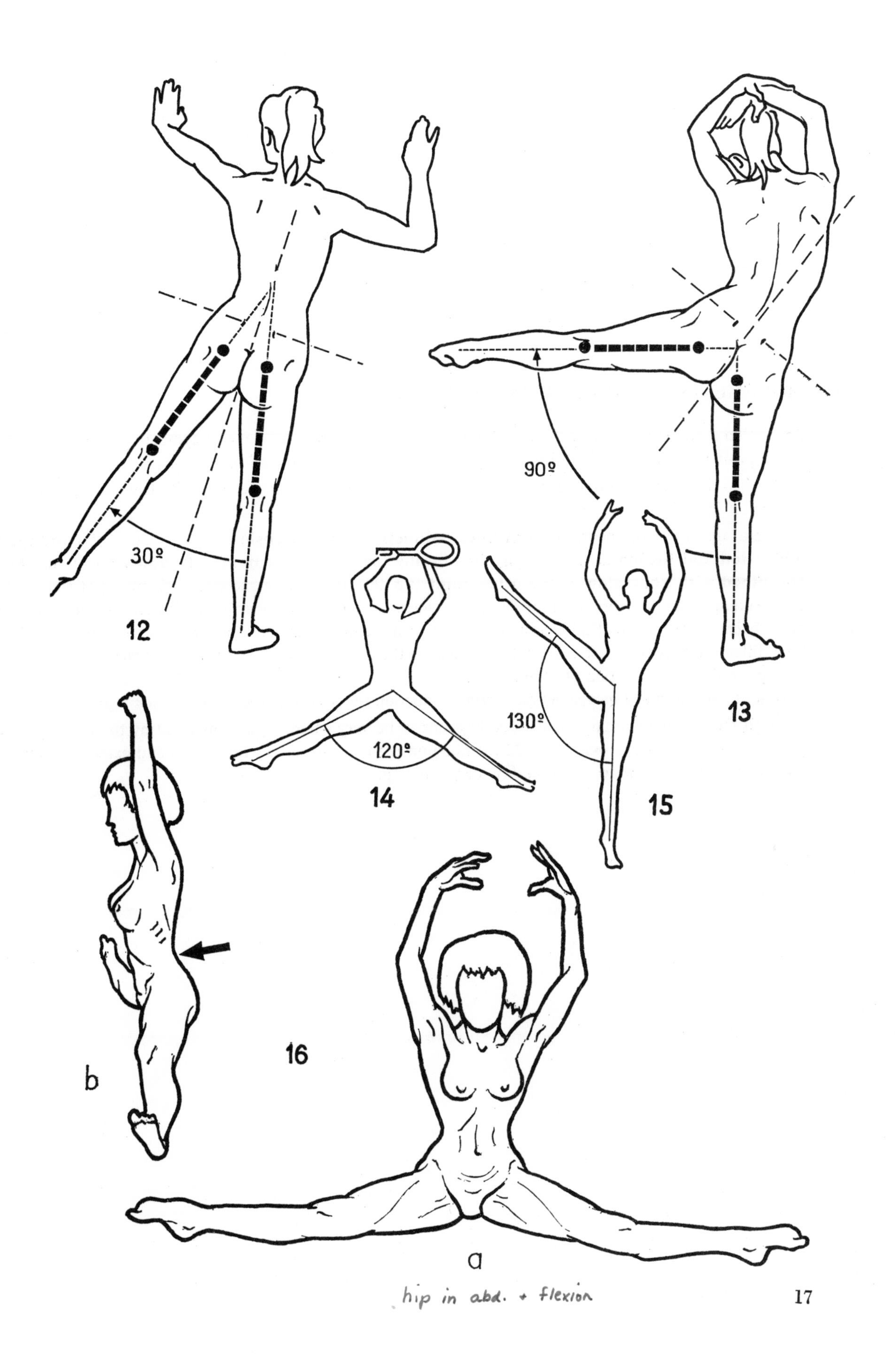

hip in abd. + flexion

MOVEMENTS OF ADDUCTION OF THE HIP

Adduction is the movement of the lower limb **medially** towards the plane of symmetry of the body. As in the position of reference both limbs are in contact, there is no *pure adduction*.

On the other hand, "**relative**" **adduction** exists as when the limb moves medially from any position of abduction (fig. 17).

There are also movements of **combined adduction and extension** (fig. 18) and of **combined adduction and flexion** at the hip (fig. 19). Finally there are **movements of adduction at one hip combined with abduction at the other joint** (fig. 20); these are associated with tilting of the pelvis and bending of the vertebral column. Note that when the feet are set apart—this is necessary to maintain one's balance—the angle of adduction at one hip is not equal to that of abduction at the other (fig. 21). The difference between these two angles is equal to the angle between the two axes of the lower limbs as they lie in the initial position of symmetry.

In all these combined movements involving adduction the *maximal range of adduction is 30°*.

Of all these combined movements, one happens to be very common, as illustrated by the position of a person sitting with legs crossed (fig. 22). Adduction is then associated with flexion and external rotation. This is the position of *maximal instability* for the hip (p. 44).

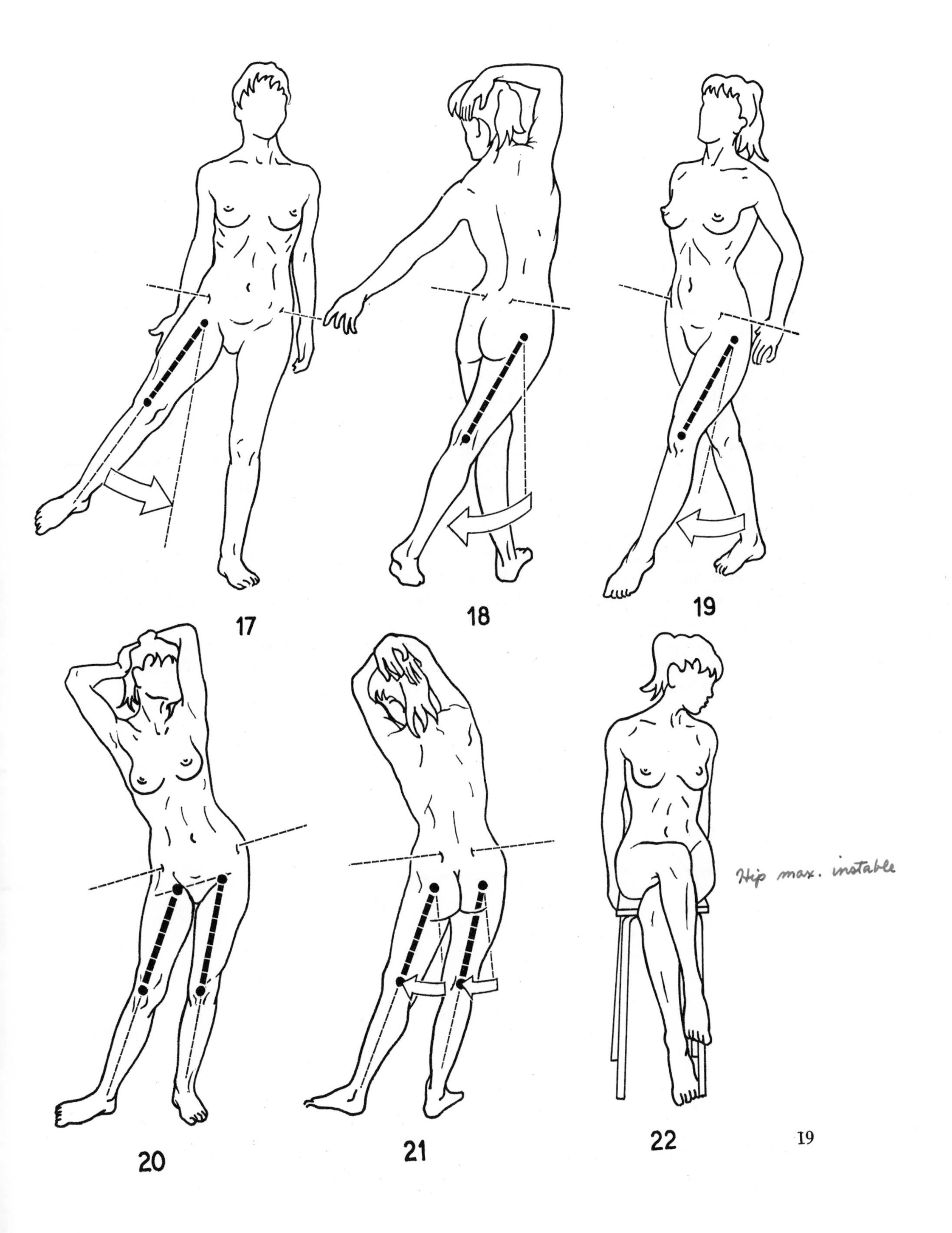
17
18
19
20
21
22
Hip max. instable

ROTATIONAL MOVEMENTS OF THE HIP

These occur about the *mechanical axis* of the lower limb (axis OR, fig. 1). In the 'straight' position this axis coincides with the vertical axis of the hip (axis OZ, fig. 1). Under these circumstances, **lateral rotation** is the movement of the limb that brings the tips of the toes to face outwards and **medial rotation** brings the tips of the toes to face inwards. As the knee is fully extended, rotation occurs only at the hip (p. 124).

However, this is not the position used for assessing the range of rotational movements. This is more easily done with the subject lying prone or sitting on the edge of a table with his knee flexed at 90°.

When the subject is *lying prone*, the **position of reference** (fig. 23) is achieved when the leg is at right angles to the thigh and is *vertical*. From this position, when the leg moves *laterally*, **medial rotation** occurs with a total range of 30° to 40°(fig. 24); when the leg moves *medially*, **lateral rotation** (fig. 25) occurs with a total range of 60°.

When the subject is *sitting on the edge of a table* with the hip and knee flexed at 90° the same criteria apply: when the leg moves medially, *lateral rotation* (fig. 26) takes place and when the leg moves laterally *medial rotation* (fig. 27) takes place. In this position the total range of lateral rotation can be *greater* than in the lying position because hip flexion relaxes the ilio- and pubo-femoral ligaments, which play a vital part in checking lateral rotation (p. 38).

In the squatting position (fig. 28) lateral rotation is combined with abduction and flexion exceeding 90°. Yoga experts can achieve such a degree of lateral rotation that the two legs become parallel and horizontal ('position of the lotus').

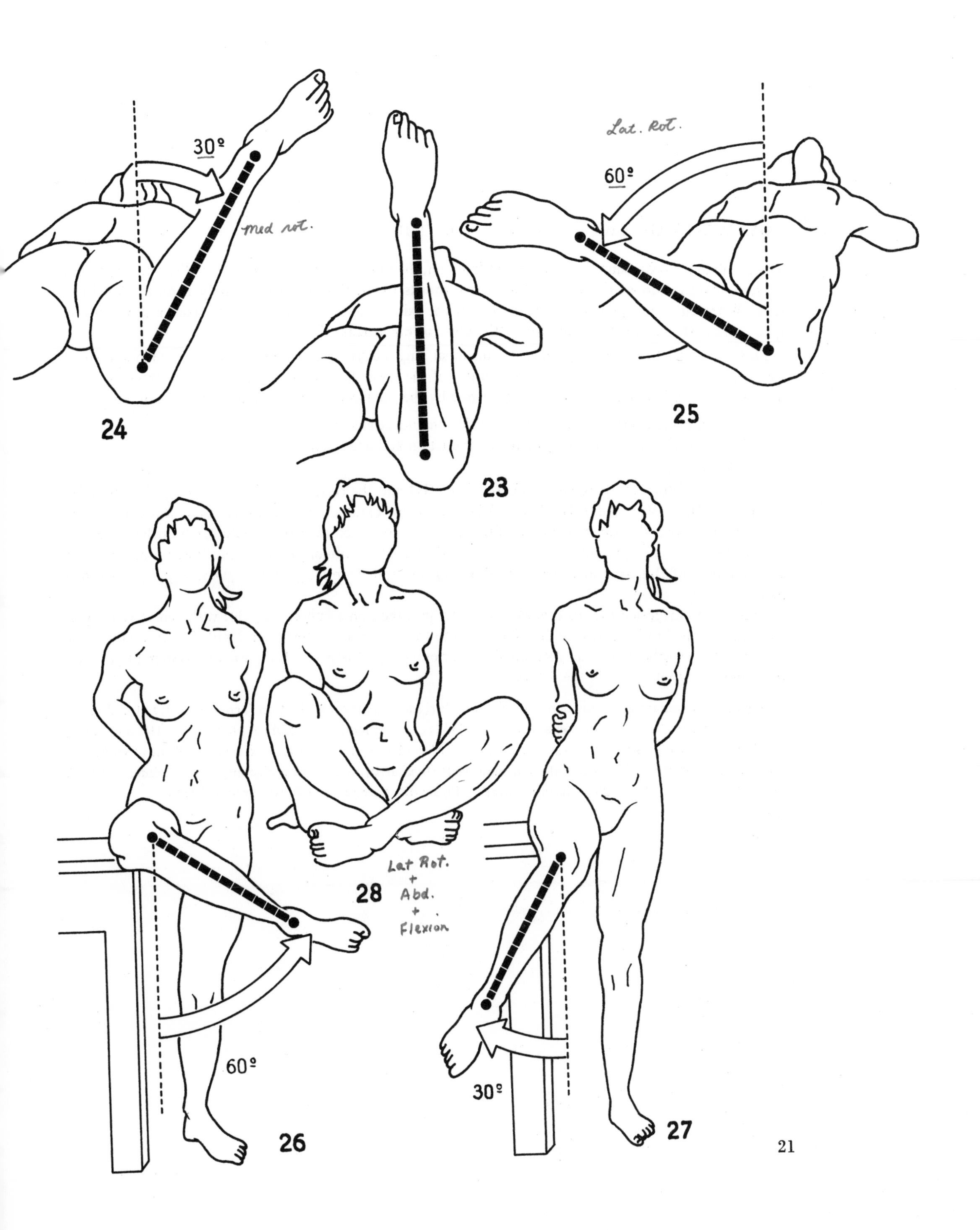
30º
med rot.
Lat. Rot.
60º
24
23
25
28
Lat Rot.
+
Abd.
+
Flexion
60º
26
30º
27

MOVEMENTS OF CIRCUMDUCTION OF THE HIP

As with all joints possessing three degrees of freedom, the movement of circumduction of the hip is defined as **the combination of the elementary movements occurring simultaneously around the three axes**. When circumduction is of maximal range, the axis of the lower limb traces in space a cone with its apex lying at the centre of the hip: this is the **cone of circumduction** (fig. 29).

This cone is far from symmetrical as the maximal ranges of the various elementary movements in space are not equal. The path traced by the extremity of the lower limb is not a circle but an irregular curve traversing the various sectors of space established by the intersection of the three planes of reference:

A. Sagittal plane containing movements of flexion and extension;

B. Frontal plane containing movements of abduction and adduction;

C. Horizontal plane.

The eight sectors of space are numbered I to VIII and the cone traverses successively the following sectors: III, II, I, IV, V and VIII. (Sector VIII lies below plane C, diagonally opposite sector IV.) Note how the curve skirts the supporting limb; if the latter were removed the curve would reach further medially. The arrow R which represents the distal, anterior and lateral prolongation of the lower limb in sector IV is the **axis of the cone of circumduction** and corresponds to *the position of function and of immobilisation of the hip*.

Strasser has suggested that this curve should be inscribed on a sphere (fig. 30) with centre O lying at the centre of the hip joint, with radius OL equal to the length of the femur and with EI representing the equator. On this sphere, one can determine the various range maxima with the use of a system of latitudes and longitudes (not shown in the diagram).

Starting from an arbitrary position OL of the femur, movements of abduction (arrow Ab) and of abduction (arrow Ad) occur along the horizontal meridian (HM); movements of medial rotation (arrow IR) and lateral rotation (arrow ER) take place about the axis OL. Movements of flexion and extension fall into two groups depending on whether they occur along a parallel P (flexion F_1, then called circumpolar) or along the large circle C (flexion F_2, then called circumcentral). These distinctions are of little practical value.

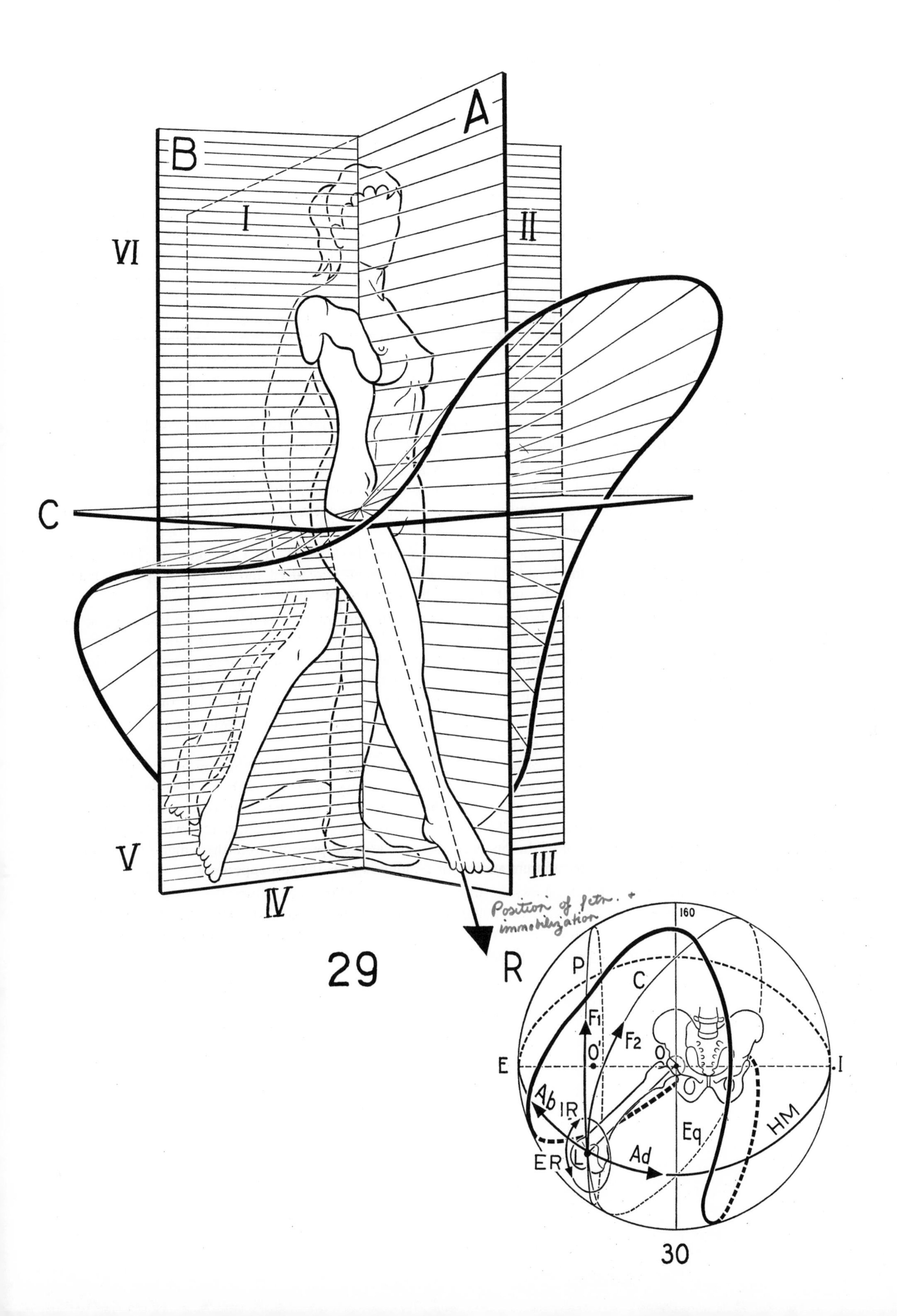

A
B
I
II
VI
C
V
III
IV
R
29
160
P
C
F1
F2
O'
O
E
I
Ab
IR
ER
L
Ad
Eq
HM
30

ARTICULAR SURFACES AND STRUCTURE OF THE HIP

(the numbers are common to all the diagrams)

The hip is of the **ball-and-socket variety** with **spherical** articular surfaces.

The **femoral head** (fig. 31, seen from in front) forms about two-thirds of a sphere of diameter 4 to 5 cm. Its geometrical centre is traversed by the three axes of the joint: horizontal axis (1), vertical axis (2), antero-posterior axis (3). The head is supported by the neck of the femur which joins the shaft. The axis of the femoral neck is obliquely set and runs superiorly, medially and anteriorly. In the adult it forms an obtuse angle of 125° with the femoral shaft (D) and an acute angle of 10° to 30° with the frontal plane (fig. 37, seen from above); this angle faces medially and anteriorly and is also called the *angle of anteversion*. Therefore (fig. 34, seen from behind and from inside) the coronal plane through the centre of the femoral head and the axis of the femoral condyles (plane P) lies almost completely *anterior* to the femoral shaft and its upper extremity. *This plane P contains the mechanical axis* MM′ of the lower limb and this axis forms an angle of 5° to 7° with the axis of the shaft (D) (p. 74).

The shape of the head and neck varies considerably with the individual and, according to anthropologists, it is the result of functional adaptation. Two extreme types are described (fig. 35, according to Bellugue):—

Type I: the head is more than two-thirds of a sphere, the angle between the neck and the shaft (I = 125°) and that between the neck and the frontal plane (D = 25°) are maximal. The shaft is slender and the pelvis is small and high slung. Such a configuration favours range of movement at the joint and corresponds to an adaptation to speed of movement (fig. 35, a and c).

Type II: the head just exceeds a hemisphere, the angles (I = 115°, D = 10°) are minimal. The shaft is thicker and the pelvis large and broad. The range of movement is reduced and the loss of speed is made up for by the greater strength of the joint. This is the configuration 'of power'.

The **acetabulum** (fig. 32; seen from outside) receives the femoral head and lies on the lateral aspect of the hip bone where its three constituent bones meet. It is hemispherical and is bounded by the acetabular rim (R). Only the sides of the acetabulum are lined by a *horseshoe-shaped articular cartilage* (Ca), which is interrupted inferiorly by the deep *acetabular notch*. The central part of the cavity is deeper than the articular cartilage and is non-articular: it is called the *acetabular fossa* (AF) and is separated from the inner face of the pelvic bone by a thin plate of bone (fig. 33; transparent bone). It will be shown later (p. 30) how the *labrum acetabulare* is applied to the acetabular rim.

The acetabulum is directed *laterally, inferiorly and anteriorly* (arrow C′ representing the axis of the acetabulum). A vertical section of the acetabulum (fig. 36) shows quite clearly that it faces inferiorly: the acetabular axis forms an angle of 30° to 40° with the horizontal, so that the upper part of the acetabulum 'overhangs' the femoral head laterally. This degree of overhanging is measured by the angle W (Wiberg), which is normally 30°. The roof of the cavity sustains the greatest pressure from the femoral head and so the articular cartilage of the acetabulum and of the femoral head is thickest superiorly.

The horizontal section (fig. 37) shows the anterior orientation of the acetabulum: the axis C′ is at an angle of 30° to 40° with the frontal plane. Also included are: the acetabular fossa (AF) lying deep to the articular cartilage (Ca); the labrum acetabulare (LA) continuous with the acetabular rim; the plane tangential to the rim (R), which runs obliquely anteriorly and medially.

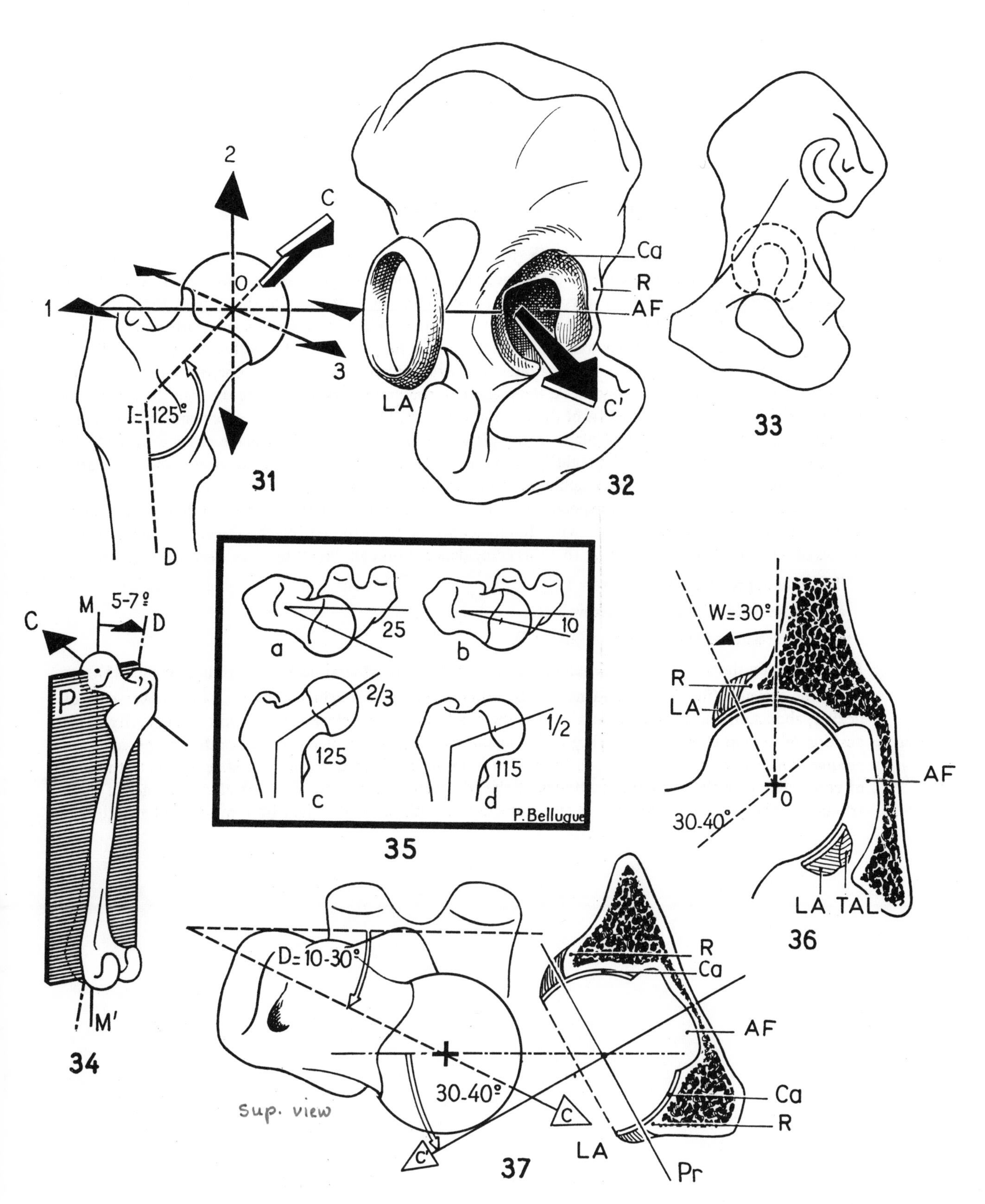

2
C
O
1
3
I= 125°
D
31
Ca
R
AF
LA
C'
32
33
M
5-7°
C
D
P
M'
34
25
a
10
b
2/3
125
c
1/2
115
d
P. Bellugue
35
W= 30°
R
LA
AF
O
30-40°
LA
TAL
36
D= 10-30°
30-40°
C
C'
R
Ca
AF
Ca
R
LA
Pr
37
sup. view

RELATIONSHIPS OF THE ARTICULAR SURFACES

When the hip is in **the 'straight' position** (fig. 38), which corresponds to *the erect posture* (fig. 39), the femoral head is not completely covered by the acetabulum, the cartilage-lined anterosuperior aspect being exposed (arrowed fig. 38). This results (fig. 44, three-dimensional diagram of the axes of the right hip) from the fact that the axis of the femoral neck (C), which runs obliquely superiorly, anteriorly and medially, is out of line with the acetabular axis (C′) which runs obliquely inferiorly, *anteriorly* and laterally. A mechanical model of the hip (fig. 40) illustrates this arrangement as follows: A sphere is fixed to a shaft so as to mimic the femoral head and neck; the plane D represents the plane passing through the axis of the femoral shaft, and the transverse axis of the femoral condyles. On the other hand, a hemisphere is suitably arranged in relation to the sagittal plane S; the plane F represents the frontal plane passing through the centre of the hemisphere. In the 'straight' position, the sphere is largely exposed superiorly and anteriorly: the black crescent represents the portion of articular cartilage which is exposed.

By moving the 'acetabular hemisphere' and the 'femoral sphere' (fig. 43) one can achieve complete coincidence of the articular surfaces with disappearance of the exposed 'black crescent'. Thanks to the planes of reference S and P, it is clear that this coincidence is brought about by *three elementary movements*:

flexion of approximately 90° (arrow 1);
a small measure of abduction (arrow 2);
a small measure of lateral rotation (arrow 3).

In this position the axis of the acetabulum (C′) and that of the femoral neck (C″) are in line (fig. 45).

On the skeleton (fig. 41), coincidence of the articular surfaces is achieved by the same movements of flexion, abduction and lateral rotation so that the head lies completely within the acetabular cavity. This position of the hip corresponds to the *position on all fours* (fig. 42), which is therefore the *true physiological position of the hip.* During evolution, the transition from the quadruped to the biped state has led to **the loss of coincidence of the articular surfaces of the hip joint.** Conversely, this lack of coincidence of these articular surfaces can be considered as an argument in favour of man's origin from quadruped ancestors.

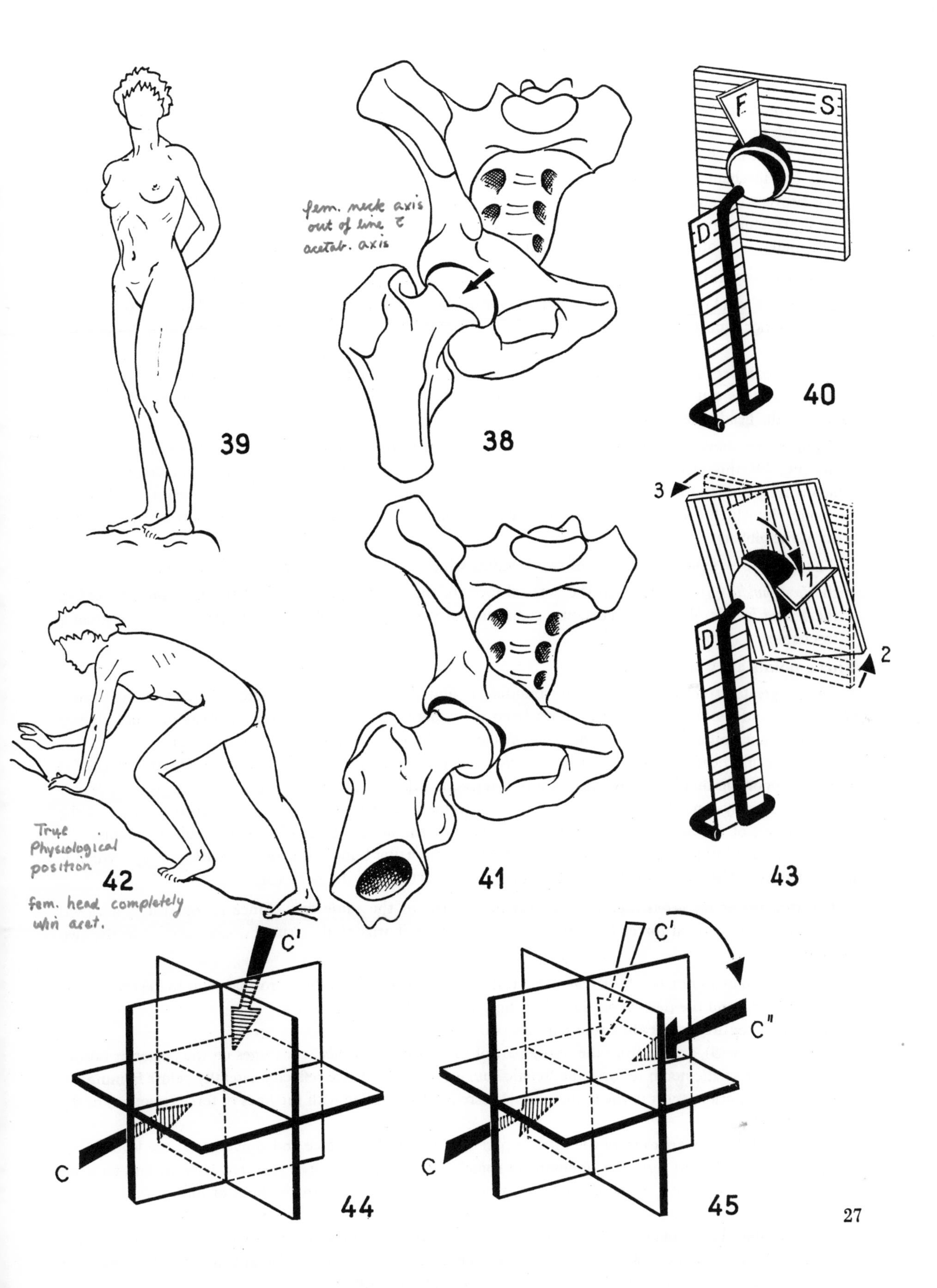
39
38
fem. neck axis out of line c̄ acetab. axis
F
S
D
40
3
1
2
43
41
True Physiological position
42
fem. head completely with acet.
C'
C
44
C'
C"
C
45

THE STRUCTURE OF THE FEMUR AND PELVIS

The head, neck and shaft of the femur can be compared to a structure known as an '*overhang*' in engineering. In fact, the weight of the body, when applied to the femoral head, is transmitted to the shaft *by a lever arm*, the femoral neck. A similar set-up is seen in the *gibbet* (fig. 50), where the weight acting vertically tends to 'shear' the horizontal beam near its junction with the shaft and so to close the angle between the two. To prevent this occurrence a *strut* is interposed obliquely.

The femoral head represents the horizontal beam of the gibbet and an overall picture (fig. 48) of the lower limb shows that the mechanical axis of its three joints (heavy broken line) runs medial to the femoral head (n.b. the mechanical axis does not coincide with the vertical shown by the line of alternate dashes and dots). The mechanical significance of this arrangement will emerge later (fig. 128).

To prevent the shearing of the base of the femoral head (fig. 51) the **upper end of the femur** has a special structural pattern which can be easily seen in a *vertical section* of the desiccated bone (fig. 46). The lamellae of spongy bone are arranged in *two systems of trabeculae* corresponding to the *lines of force*.

The **main system** consists of two sets of trabeculae fanning out into the neck and head:

1. the *first set* (1) arises from the cortical layer of the lateral aspect of the femoral shaft and terminates on the inferior aspect of the cortical layer of the femoral head (the so-called *arcuate bundle* of Gallois and Bosquette).

2. the *second set* (2), arising from the cortex of the internal aspect of the shaft and of the inferior part of the neck, fans out vertically in an upward direction to terminate on the cortical bone of the superior aspect of the head (the so-called '*supporting bundle*').

Culmann has shown experimentally that, when a test bar is loaded and bent into the shape of a crook or a crane (fig. 49), this gives rise to two sets of lines of force: an oblique set, appearing on the convex aspect, which corresponds to the *shearing stresses* and is the counterpart of the arcuate bundle; a vertical set, lying in the concavity which corresponds to *compressive forces* and is the counterpart of the 'supporting bundle' (the strut of the gibbet).

The **accessory system**, consists of two bundles which fan out into the greater trochanter:

the *first bundle* (3), arising from the cortical layer of the inner aspect of the shaft (*trochanteric bundle*);
the *second bundle* (4) (less important), consisting of vertical trabeculae running parallel to the greater trochanter.

Three points are worth noting:

1. In the greater trochanter the arcuate bundle (1) and the trochanteric bundle (3) intersect to form a *Gothic arch* and its *keystone*, running down from the superior aspect of the neck, consists of much denser bone. The inner pillar (3) is less strong and weakens with age as a result of senile osteoporosis.
2. The neck and head also contain *another Gothic arch* formed by the intersection of the arcuate bundle (1) and the supporting bundle (2). At the point of intersection the bone is denser and constitutes the '*nucleus of the head*'. This system of trabeculae rests on an extremely strong support, the thick cortical layer of the inferior aspect of the neck, known as the inferior spur of the neck (SP) or else as the vault of Adams.
3. Between the Gothic arch of the trochanter and the supporting bundle is a *zone of weakness* (±), which is intensified by senile osteoporosis: it is the site of basal fractures of the neck (fig. 51).

The *structure of the pelvis* (fig. 46) can also be studied in the same way. Since it constitutes a closed ring, it transmits vertical forces from the vertebral column (horizontally striped double arrow) to the two hip joints.

There are two main *systems of trabeculae* that transmit the stresses from the sacro-iliac joint to the *acetabulum* on the one hand and the *ischium* on the other (figs. 46 and 47).

The **sacro-acetabular trabeculae** fall into two sets:

1. The first set (5), arising from the upper part of the auricular surface, converges on the posterior border of the greater sciatic notch to form the 'sciatic spur' (SS). It is thence reflected laterally before fanning out towards the inferior aspect of the acetabulum, where it falls into line with the lines of force (traction) of the femoral neck (1).
2. The second set (6) arises from the inferior part of the auricular surface and converges at the level of the superior gluteal line to form the 'innominate spine' (IS). From there it is reflected laterally and fans out towards the upper aspect of the acetabulum, where it falls into line with the lines of force (compression) of the femoral neck (2).

The **sacro-ischial trabeculae** (7) arise from the auricular surface in conjunction with the above-mentioned bundles and then run downwards to the ischium. They intersect the trabeculae arising from the acetabular rim (8). These trabeculae bear the body weight in the sitting position.

Finally, the trabeculae arising from the innominate spine (IS) and the 'sciatic spur' run together into the horizontal ramus of the pubis to complete the pelvic ring.

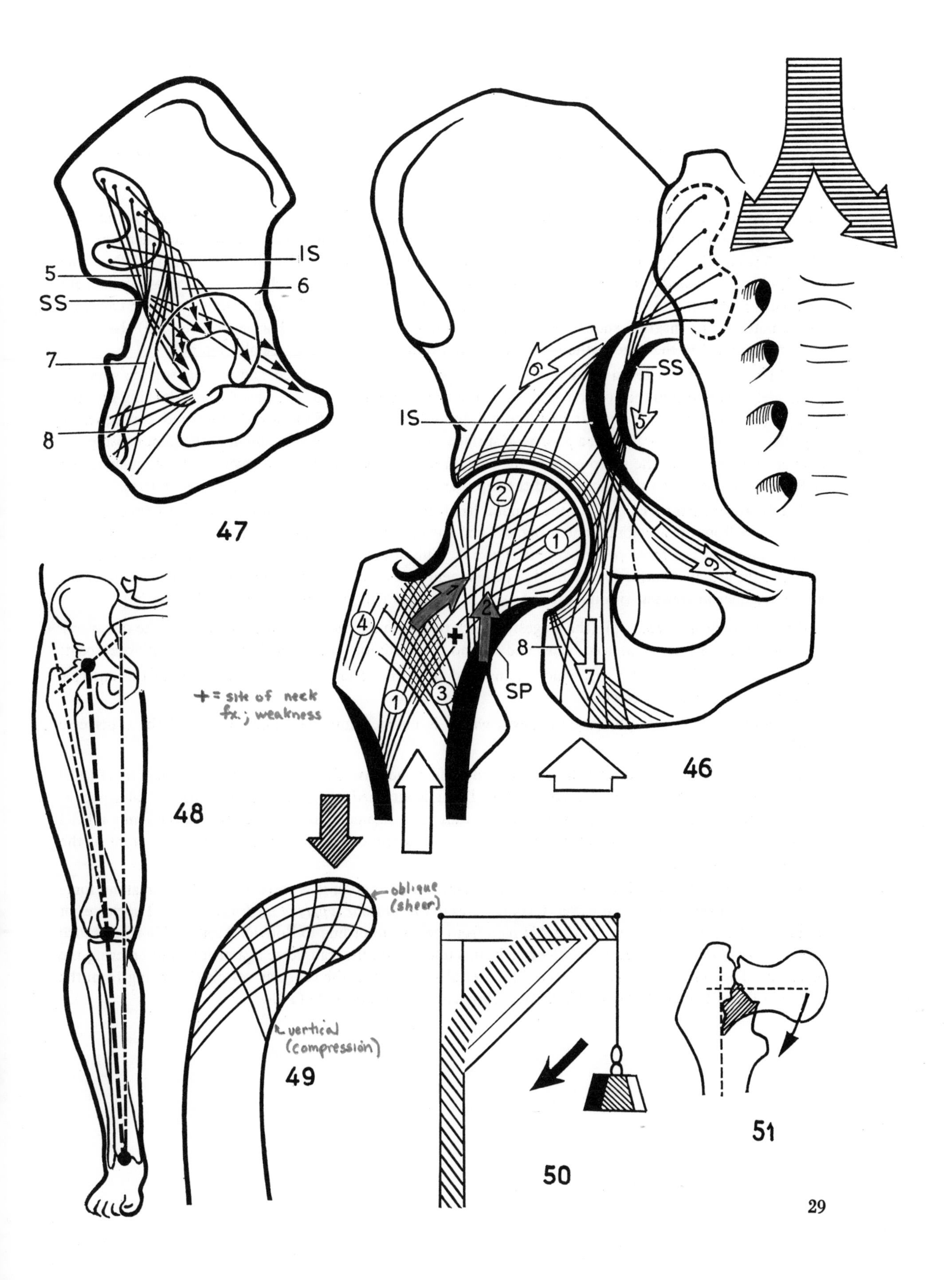
IS
5
6
SS
7
8
47
6
IS
SS
5
2
1
9
4
2
8
SP
1
3
7
+ = site of neck fx.; weakness
46
48
oblique (shear)
vertical (compression)
49
50
51

THE LABRUM ACETABULARE AND THE LIGAMENTUM TERES OF THE FEMORAL HEAD

The **labrum acetabulare** (LA) is a **fibrocartilaginous ring** inserted into the acetabular rim (fig. 52). It deepens the acetabulum considerably (p. 44) and fills out the various gaps of the acetabular rim (R). The anterosuperior aspect of the labrum has been removed and shows the 'iliopubic notch' (IPN). The ischiopubic or *acetabular notch* (AN), which is the deepest of the three notches, is bridged by the labrum as it gains insertion into the *transverse acetabular ligament* (TAL) which is itself inserted into the two sides of the notch. (The diagram shows the ligament and labrum 'displaced'.) The section (fig. 53) shows the labrum well fixed to the edge of the notch and to the transverse ligament (see also fig. 36).

The labrum is in fact **triangular** on section and possesses **three surfaces**: an *internal* surface which is completely inserted into the acetabular rim and the transverse ligament, a *central* surface (looking into the joint), which is lined by articular cartilage continuous with that of the acetabulum and consequently articulates with the femoral head; a *peripheral* surface which receives the attachment of the joint capsule (C) only at its base so that the sharp edge of the labrum lies free within the joint cavity and a circular recess (CR) is formed between the labrum and the capsular attachment (fig. 54, according to Rouvière).

The **ligamentum teres** (LT) of the femoral head (fig. 56) is a flattened fibrous band 3 to 3·5 cm. long, which arises from the acetabular notch (fig. 52) and runs on the floor of the acetabular fossa (fig. 53) before its insertion into the *fovea femoris capitis* (fig. 55). This fovea lies slightly inferior and posterior to the centre of the articular surface of the head. The ligament is inserted into the anterosuperior part of the fovea and only glides in contact with its inferior surface. The ligament consists of *three bundles* (fig. 56):

the *posterior ischial bundle* (pi) (the longest), which runs through the acetabular notch under the transverse ligament (fig. 52) to be inserted below and behind the posterior horn of the horseshoe articular crescent;

the *anterior pubic bundle* (ap), which is inserted into the acetabular notch itself behind the anterior horn of the articular crescent;

the *intermediate bundle* (ib) (the thinnest), which is inserted into the upper border of the transverse ligament (fig. 52).

The ligamentum teres (fig. 53) lies embedded in fibro-adipose tissue within the acetabular fossa (AF) and is lined by synovium (fig. 54). The synovial lining is attached on the one hand to the central aspect of the articular crescent and the upper aspect of the transverse ligament, and on the other to the femoral head around the fovea. It therefore has roughly the shape of a truncated cone, hence its name of 'tent of the ligamentum teres' (Ts).

The ligamentum teres plays only a trivial mechanical role though it is extremely strong (breaking force equivalent to 45 kg. weight). However, it contributes to the *vascular supply of the femoral head*. The posterior branch of the obturator artery (1) (fig. 57: seen from below, according to Rouvière) sends off a tiny branch —*the artery of the ligamentum teres* (6)—which runs underneath the transverse ligament before entering the ligament. The head and neck also receive an arterial supply from the capsular vessels (5) arising from the medial (3) and lateral (4) circumflex arteries (branches of the profunda (2)).

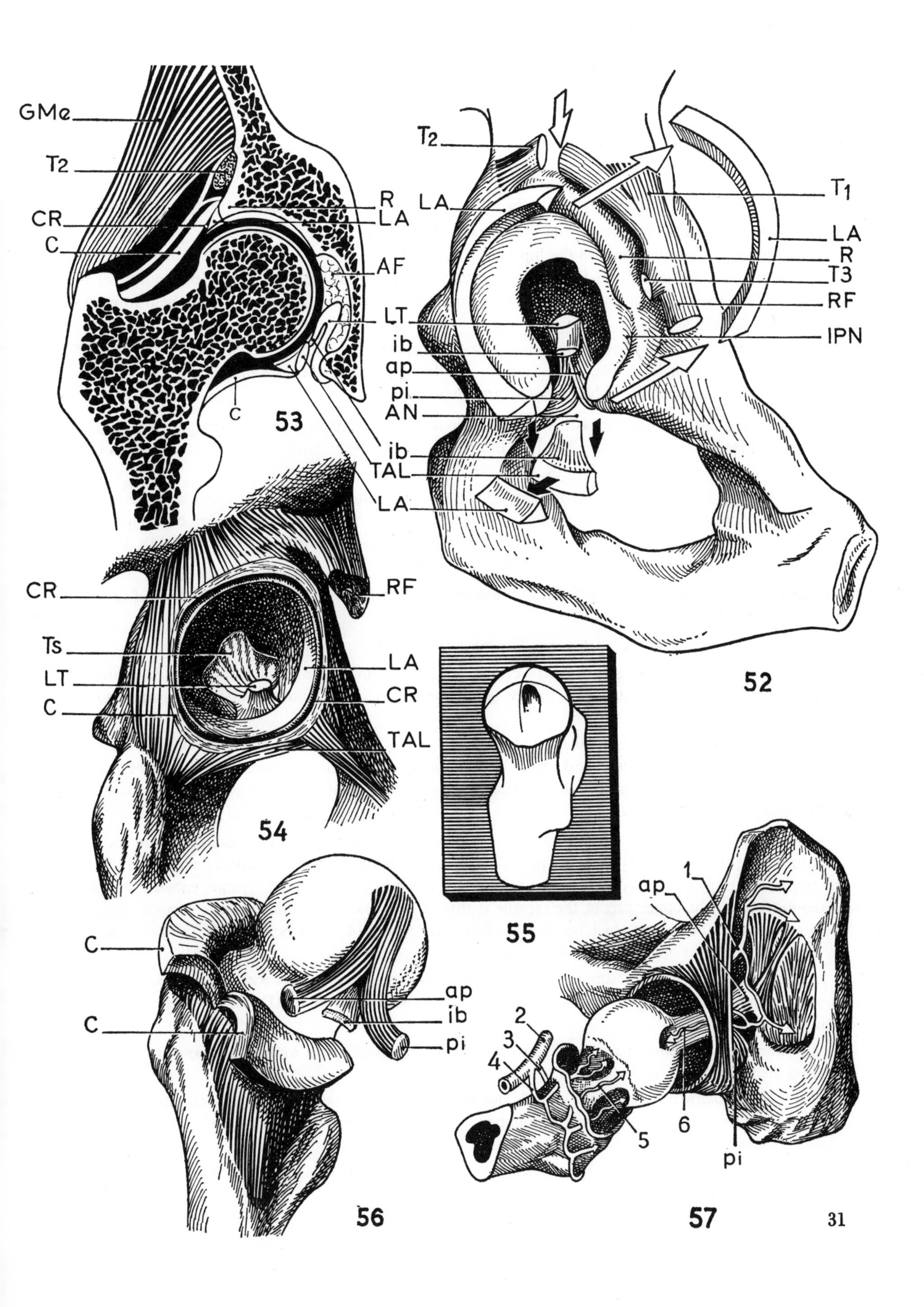

GMe
T2
CR
C
R
LA
AF
LT
c
53
ib
TAL
LA
T2
LA
T1
LA
R
T3
RF
IPN
LT
ib
ap
pi
AN
ib
TAL
LA
52
CR
RF
Ts
LT
C
LA
CR
TAL
54
55
C
C
ap
ib
pi
56
1
ap
2
3
4
5
6
pi
57

THE CAPSULE AND LIGAMENTS OF THE HIP

THE CAPSULAR LIGAMENT OF THE HIP

The capsule is shaped like a **cylindrical sleeve,** (fig. 58) running from the iliac bone to the upper end of the femur. It is made up of *four distinct sets of fibres*:

longitudinal fibres (1), which help to unite the articular surfaces and run parallel to the axis of the cylinder;
oblique fibres (2), of similar function to that of (1), form a spiral round the cylinder;
arcuate fibres (3), attached only to the hip bone. They run in a criss-cross fashion from one end of the acetabular rim to the other and form an arc of varying height and apex flush with the middle of the sleeve. These arcuate fibres are arranged like a man's tie around the femoral head and help to keep it within the acetabulum.

circular fibres (4) with no bony attachments. They are particularly abundant in the middle of the sleeve which they groove slightly. They stand out on the deep surface of the capsule and are known as the *zona orbicularis* (the ring of Weber), which surrounds the neck.

Medially, the capsular ligament is inserted into the acetabular rim (5), the transverse ligament and the peripheral surface of the labrum (p. 30). It is intimately related to the tendon of the rectus femoris (RF, fig. 52) as follows:—

The *straight head* (T_1) of the rectus femoris arising from the antero-inferior iliac spine, and the *reflected head* (T_2), arising from the groove above the rim of the acetabulum, unite before running between the two slips of the capsular insertion (fig. 53) reinforced superiorly by the ilio-femoral ligament (p. 34). The *deep recurrent fibres* (T_3) strengthen the anterior as pect of the capsule.

Laterally, the capsule is not inserted into the edges of the articular cartilage but at the base of the neck along a line which runs:

anteriorly (fig. 58), along *the trochanteric line* (6);
posteriorly (fig. 59), not along trochanteric crest (7), but at the junction of the lateral and middle thirds of the femoral neck (8), just above the *groove* (9) of the obturator externus before its insertion into the trochanteric fossa (TF).
The line of insertion of the capsule is oblique to both the inferior and superior surfaces of the neck. Inferiorly (10), it runs superior and about 1·5 cm. anterior to the lesser trochanter (Lt). The deepest fibres extend up the lower surface of the neck to reach the edge of the cartilage and in so doing they raise synovial folds (*frenula capsulae* (11)), the longest of which is the *pectinofoveal fold of Amantini* (12).

These frenula capsulae are useful during movements of abduction.. During adduction (fig. 60) the inferior part of the capsule (1) slackens while the upper part becomes taut (2). During abduction (fig. 61) the frenula (3) unpleat and, by increasing the length of the inferior part of the capsule, enhance the range of the movement: the upper part of the capsule is thrown into folds (2) while the neck impacts on to the acetabular rim via the labrum which becomes distorted and everted (4). This explains why the *labrum deepens the acetabulum without limiting movements at the joint.*

In extreme flexion the anterosuperior aspect of the neck comes into contact with the acetabular rim and in some individuals (fig. 58) the neck at this point bears an iliac impression (II) just above the edge of the articular cartilage.

After injection of a radio-opaque medium into the hip, skiagrams (**hip arthrograms**) can be obtained showing the following features (fig. 62):

The *zona orbicularis* (9) indents the capsule distinctly in the middle and divides the joint cavity into two chambers: a *lateral* (1) and *a medial* (2) chamber. These two chambers form the *superior recesses* (3) above and the *inferior recesses* (4) below. The medial chamber also contains:

above, a spur-like recess with its apex pointing towards the acetabular rim, the so-called *supralimbic recess* (5) (compare with fig. 53);

below, two rounded peninsulae separated by a deep gulf: these are respectively the *two acetabular recesses* (6) and the capsular impression of the *ligamentum teres* (7).

Finally can be seen the *interspace* (8) between the femoral head and the acetabulum.

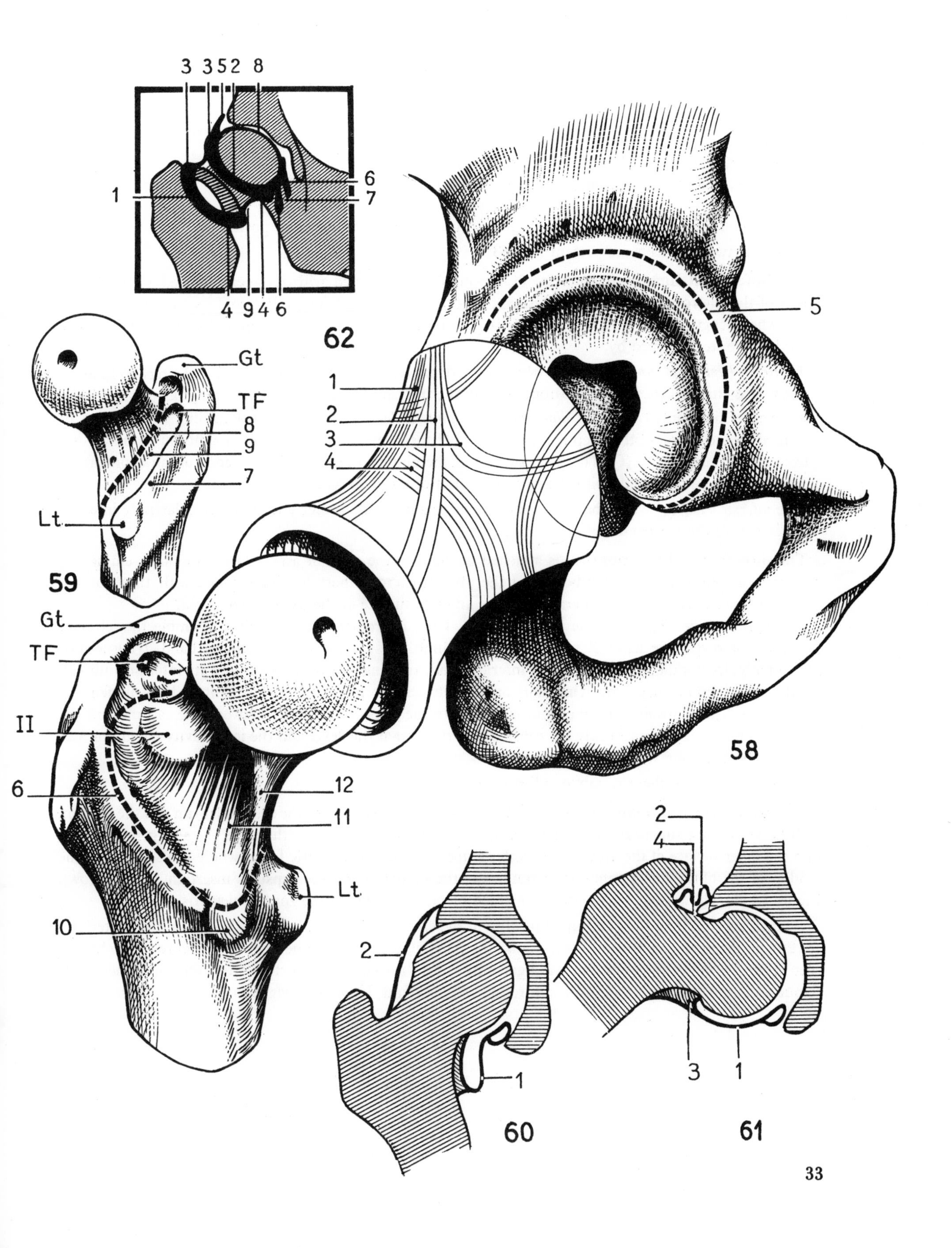
3 3 5 2 8
1
6
7
4 9 4 6
62
Gt
TF
8
9
7
Lt.
59
1
2
3
4
5
Gt
TF
II
6
12
11
Lt
10
58
2
4
2
1
3
1
60
61

THE LIGAMENT OF THE HIP

(the numbers refer to the same structures in all the diagrams)

The capsule of the hip is strengthened by powerful ligaments anteriorly and posteriorly.

Anteriorly two ligaments are present (fig. 63):

the **iliofemoral ligament** (ligament of Bertin) (1), fan-shaped with its apex attached to the lower part of the antero-inferior iliac spine (site of origin of the rectus femoris, RF) and its base inserted into the whole length of the trochanteric line. Its central part (c) is relatively thin and weak while its two borders are strengthened by:

the *iliotrochanteric or superior band* (a) which is the strongest of the ligaments of the joint, being 8 to 10 mm. thick. It is attached laterally to the upper part of the trochanteric line. It is itself strengthened superiorly by another ligament, called the *ilio-tendino-trochanteric ligament* (d), which according to Rouvière is formed by the fusion of the deep recurrent fibres of the rectus femoris (e) and of a fibrous band arising from the acetabular rim (f). The deep surface of the gluteus minimus (GMi) sends off an aponeurotic expansion (g) which blends with the external aspect of the iliotrochanteric ligament.

The *inferior band* (b) which has the same site of origin as the former and is inserted laterally into the lower part of the trochanteric line.

The **pubofemoral ligament** (2) is attached medially to the anterior aspect of the iliopubic eminence, the superior ramus of the pubic bone and the obturator crest where its fibres blend with those of the pectineus muscle. It is inserted laterally into the anterior surface of the trochanteric fossa.

Taken as a whole (fig. 64), these two ligaments, lying in front of the hip joint, resemble the letter N (Welcker) or better the letter Z with its superior limb (a), i.e. the iliotrochanteric band, lying almost horizontally, its middle limb (b), i.e. the inferior band, running nearly vertically and its inferior limb (2), i.e. the pubofermoral ligament, lying horizontal. Between the pubofemoral ligament and the iliofemoral ligament (+), the capsule is thinner and is related to the bursa intervening between the capsule and the iliopsoas tendon (IP). Occasionally the capsule is perforated at this level and the joint cavity communicates with the iliopsoas bursa.

Posteriorly (fig. 65) there is only one ligament:

The **ischiofemoral ligament** (3) arises from the posterior surface of the acetabular rim and the labrum. Its fibres, running superiorly and laterally, cross the posterior aspect of the neck (h) and gain insertion into the inner surface of the greater trochanter anterior to the trochanteric fossa, where is also inserted the tendon of the obturator externus after traversing the groove lining the capsular insertion (white arrow). Figure 66 also shows some of its fibres (i) which blend directly with the zona orbicularis (j).

As man evolved from the quadruped posture to the erect posture and the pelvis became tilted posteriorly (p. 26), all the ligaments became *coiled* round the femoral neck in the same direction. Figure 67 (right hip, seen from the outside) shows that the ligaments run in a clockwise direction from the hip bone to the femur, i.e. *extension winds these ligaments* round the neck and *flexion unwinds them*.

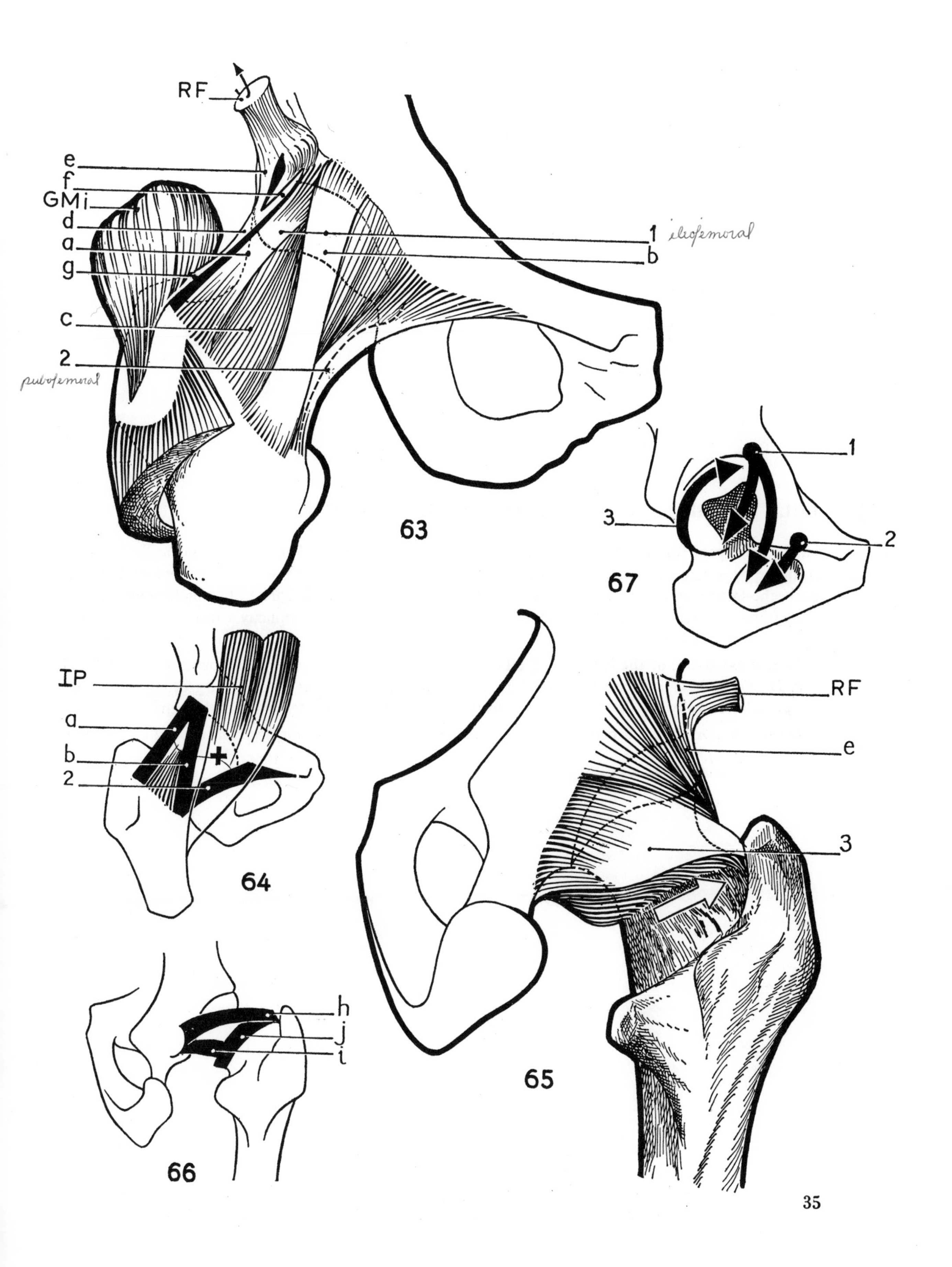
RF
e
f
GMi
d
a
g
c
2
pubofemoral
1
iliofemoral
b
63
1
3
2
67
IP
a
b
2
64
RF
e
3
65
h
j
i
66

ROLE OF THE LIGAMENTS IN FLEXION-EXTENSION

In **the erect position** (fig. 68) the ligaments are under *moderate tension.* This is diagrammatically illustrated in figure 69, where the ring represents the acetabulum and the circle in the centre the femoral head and neck, the ligaments, drawn in as springs, run between the ring and the circle. The iliofemoral ligament (ILF) and the ischiofemoral ligament (ISF) are also included (for simplicity's sake the pubofemoral ligament is not included).

During **extension of the hip** (fig. 70) all the *ligaments become taut* as they wind round the femoral neck (fig. 71). Of all these ligaments the *inferior band of the iliofemoral ligament* is under the greatest tension as it runs nearly vertically (fig. 70) and so is responsible for checking the posterior tilt of the pelvis.

During **flexion of the hip** (fig. 72) the opposite holds good and *all the ligaments without exception are relaxed* (fig. 73).

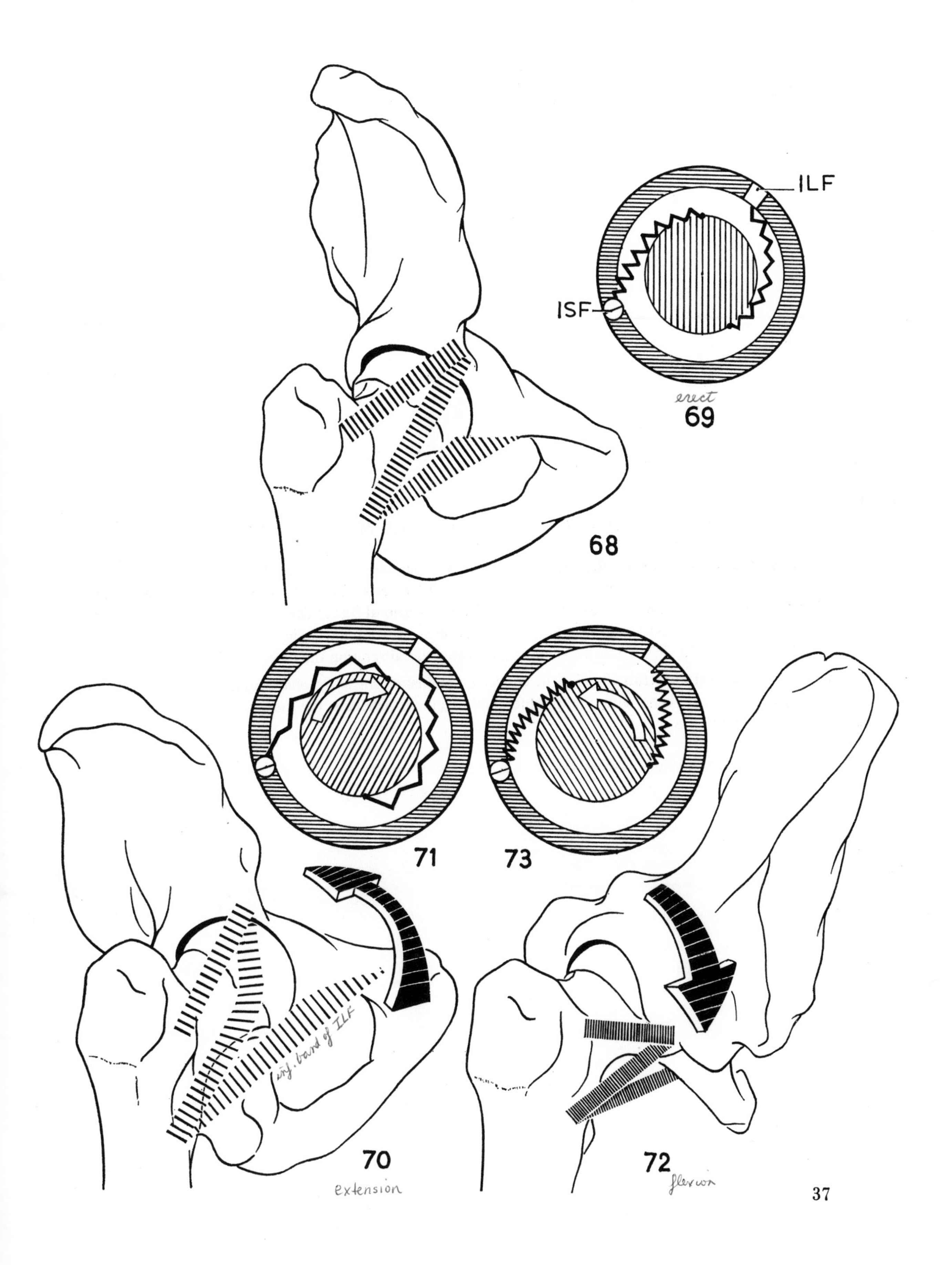
ILF
ISF
erect
69
68
71
73
inf. band of ILF
70
extension
72
flexion

ROLE OF THE LIGAMENTS IN LATERAL AND MEDIAL ROTATION

During **lateral rotation of the hip** (fig. 74) the trochanteric line moves away from the acetabular rim with the result that *all the anterior ligaments of the hip become taut* and especially those bands running horizontally, i.e. the **iliotrochanteric band and pubofemoral ligament**. This tensing up of the anterior ligaments is well demonstrated in a horizontal section seen from above (fig. 75) and in a postero-superior view of the joint (fig. 76). These also show that during lateral rotation the *ischiofemoral ligament is slackened.*

During **medial rotation** (fig. 77) the converse obtains: *all the anterior ligaments become slack,* especially the iliotrochanteric band and the pubofemoral ligament whereas the *ischiofemoral ligament tenses up* (figs. 78 and 79).

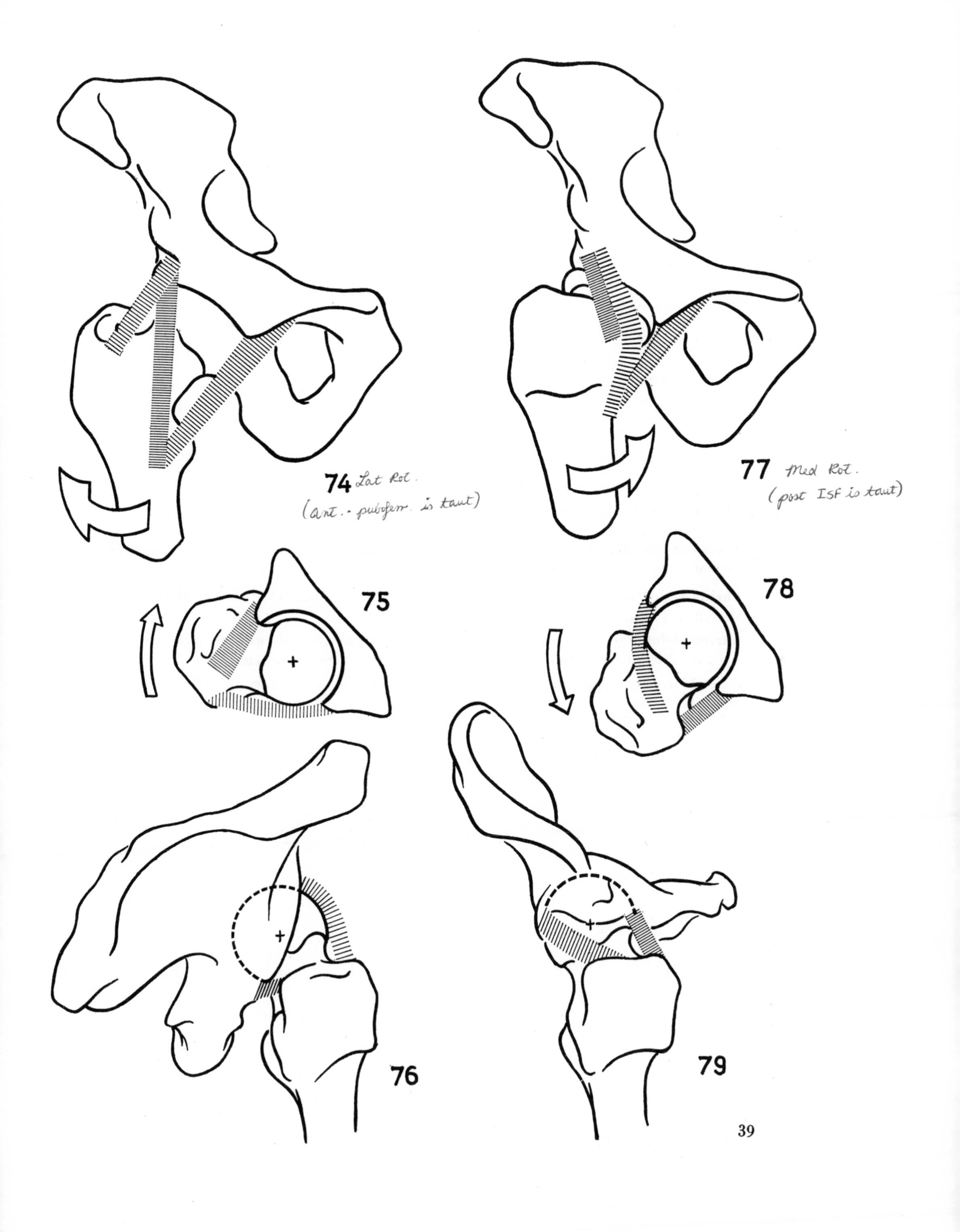
74 Lat Rot.
(ant.-pubofem. is taut)
77 Med Rot.
(post ISF is taut)
75
78
76
79

ROLE OF THE LIGAMENTS IN ADDUCTION AND ABDUCTION

Starting from the erect position (fig. 80), where the anterior ligaments are under moderate tension, it is clear that:

during adduction (fig. 81) the iliotrochanteric band becomes taut while the pubofemoral ligament is slackened. The inferior band tenses up only slightly;

during abduction (fig. 82) the opposite takes place: the pubofemoral ligaments tense up considerably while the iliotrochanteric band and, to a lesser extent, the inferior band relax;

the **ischiofemoral ligament** (seen from behind) *slackens during adduction* (fig. 83) and *tenses up during abduction* (fig. 84).

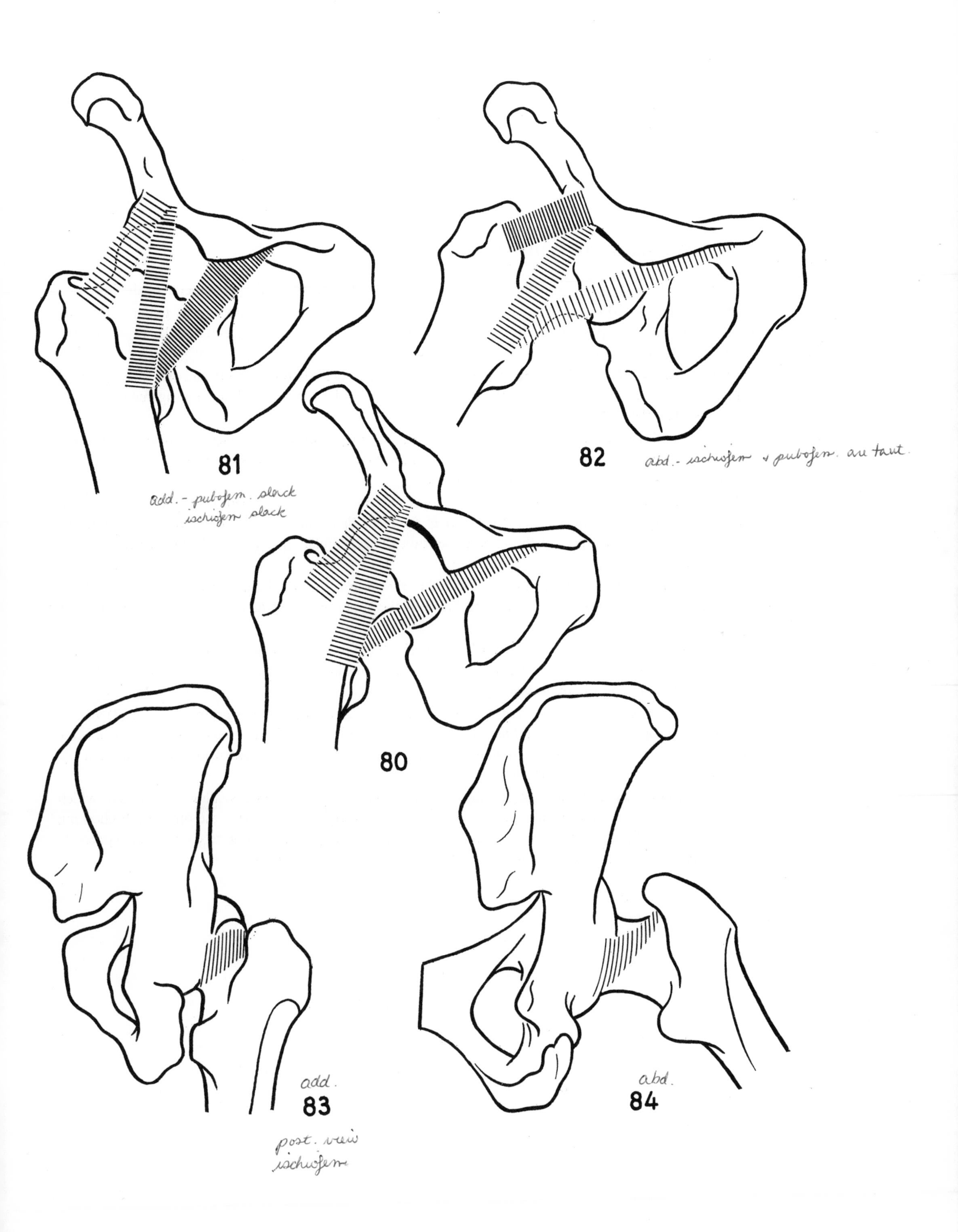
81
add. - pubofem. slack
ischiofem slack
82
abd. - ischiofem + pubofem. are taut.
80
add.
83
post. view
ischiofem
abd.
84

THE PHYSIOLOGY OF THE LIGAMENTUM TERES

This is an *anatomical vestige* and plays only a minor role in the control of the movements of the hip.

In **the erect position** (fig. 85: coronal section) it is under moderate tension and its femoral insertion lies in its intermediate position (1) in the acetabular fossa (fig. 86: diagram of the acetabular fossa showing the various positions of the fovea capitis femoris), i.e. slightly inferiorly and posteriorly to the centre (+).

During **flexion of the hip joint** (fig. 87) the ligament is twisted round itself and the fovea (fig. 86) comes to lie superior and anterior to the centre of the acetabular fossa (2). Hence the ligament plays no part in limiting flexion.

During **medial rotation** (fig. 88: horizontal section, seen from above) the fovea is displaced posteriorly and the femoral insertion of the ligament comes into contact with the posterior part of the articular crescent (3). The ligament remains moderately taut.

During **lateral rotation** (fig. 89) the fovea moves anteriorly and the ligament comes into contact with the anterior part of the articular crescent (4); here again the ligament is only moderately tensed. Note the impact of the posterior aspect of the femoral neck on the acetabular rim via the labrum, which becomes flattened and everted.

During **abduction** (fig. 90) the fovea moves inferiorly towards the acetabular notch (5) and the ligament is folded on itself. The labrum is squashed between the superior aspect of the neck and the acetabular margin.

During **adduction** (fig. 91) the fovea moves superiorly (6) to touch the roof of the acetabular fossa (6). This is the only position where the ligament is really under tension. The inferior border of the neck flattens slightly the labrum and the transverse ligament.

Therefore it becomes apparent that the acetabular fossa (including its posterior extension (7) and its anterior extension (8)) *encompasses all the various positions assumed by the foveal attachment of the ligamentum teres*. These two extensions correspond to the foveal position during adduction-extension-medial rotation (7), and adduction-flexion-lateral rotation (8).

Between these two extensions of the fossa the articular cartilage shows a shallow indentation, which corresponds to the position of minimal adduction, i.e. in the frontal plane with the other limb checking adduction early. Therefore the inner outline of the articular cartilage is not due to chance but represents *the locus of the extreme positions of the foveal attachment of the ligamentum teres.*

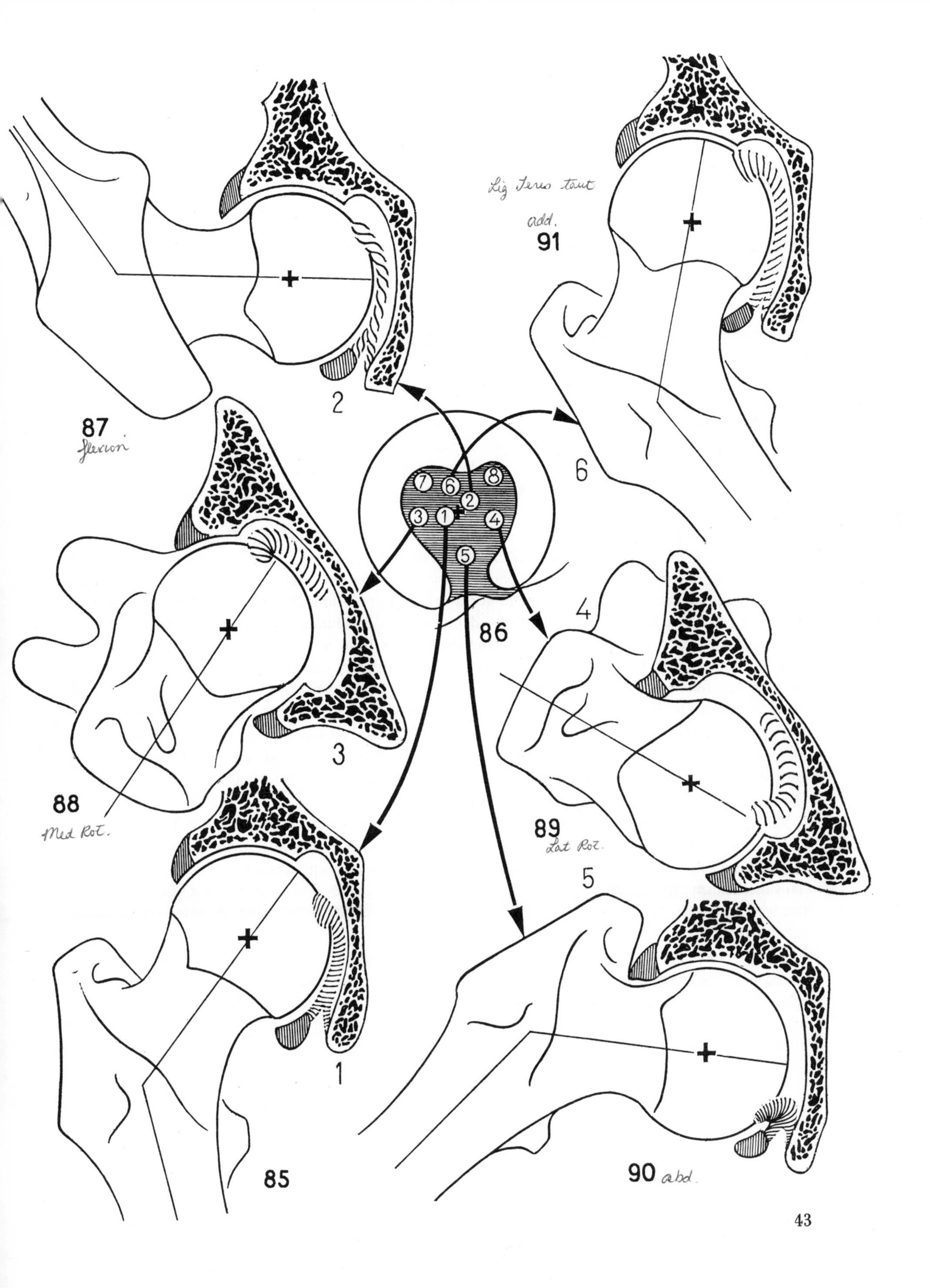

Lig Teres taut
add.
91
2
87
flexion
6
7
6
8
2
3
1
4
5
86
4
3
88
Med Rot.
89
Lat Rot.
5
1
85
90 abd.

COAPTATION OF THE ARTICULAR SURFACES OF THE HIP

In contrast with the shoulder, which tends to be dislocated by the force of gravity, the hip is assisted by **gravity**, at least in the erect position (fig. 92). To the extent that the roof of the acetabulum covers the femoral head, the latter is pressed against the acetabulum by a force (ascending arrow) equal and opposite to the weight of the body (descending arrow).

It is known that the acetabulum is not more than a hemisphere. Therefore, in mechanical terms, there cannot be *true interlocking* of the surfaces since the femoral head cannot be retained mechanically by the hemispherical bony acetabulum. However, the **labrum acetabulare** widens and deepens the acetabulum so that the *acetabular cavity exceeds a hemisphere* (black arrows). Hence the hip is transformed into a proper ball-and-socket joint with *the fibrocartilaginous labrum holding the femoral head*. This fibrous interlocking is further enhanced by the **zona orbicularis** of the capsule which encircles the femoral head (shown in section by the small arrows).

Atmospheric pressure plays an important part in maintaining apposition of the articular surfaces, as proved by *the experiments of the Weber brothers*. They noted that, if all the soft tissue connections (including the capsule) were severed between the hip bone and the femur, the femoral head did not leave the acetabulum spontaneously and in fact could only be pulled away with great difficulty (fig. 93). If on the other hand (fig. 94) a small hole had been drilled into the depths of the acetabulum, the femur fell away under its own weight. If the hole was stopped after replacing the femoral head into the acetabulum, it was again very difficult to remove the head from the acetabulum. This experiment can be compared to the *classical experiment of Magdebourg*. He showed that it is impossible to separate two hemispheres after a vacuum has been created inside (fig. 95), whereas it is very easy to do so once air has been allowed in by a tap (fig. 96).

The **ligaments** and the **periarticular muscles** play a *vital part* in the maintenance of the structural integrity of the joint. Note (fig. 97: horizontal section) that their functions are reciprocally balanced. Thus anteriorly the muscles are very few (white arrow A) and the ligaments powerful while posteriorly the muscles (B) predominate.

Note also that the action of the ligaments varies *according to the position of the hip*: in the erect position or in extension (fig. 98) the ligaments are under tension and are efficient in securing coaptation; in flexion (fig. 99) the ligaments are relaxed (p. 36) and the femoral head is not as powerfully applied to the acetabulum. This mechanism can be understood easily from the mechanical model (fig. 100): parallel fibres run between two wooden circles (a) and, when one circle moves circularly relative to the other (b), the distance between them is reduced.

The *position of flexion* is therefore a *position of instability* because of the slackness of the ligaments. When a measure of adduction is added to the flexion, as in the sitting position with legs crossed (fig. 101), a relatively mild force applied along the femoral axis (arrow) is enough to cause posterior dislocation of the hip joint with or without fracture of the posterior margin of the acetabulum (e.g. impact on the dashboard during car accidents).

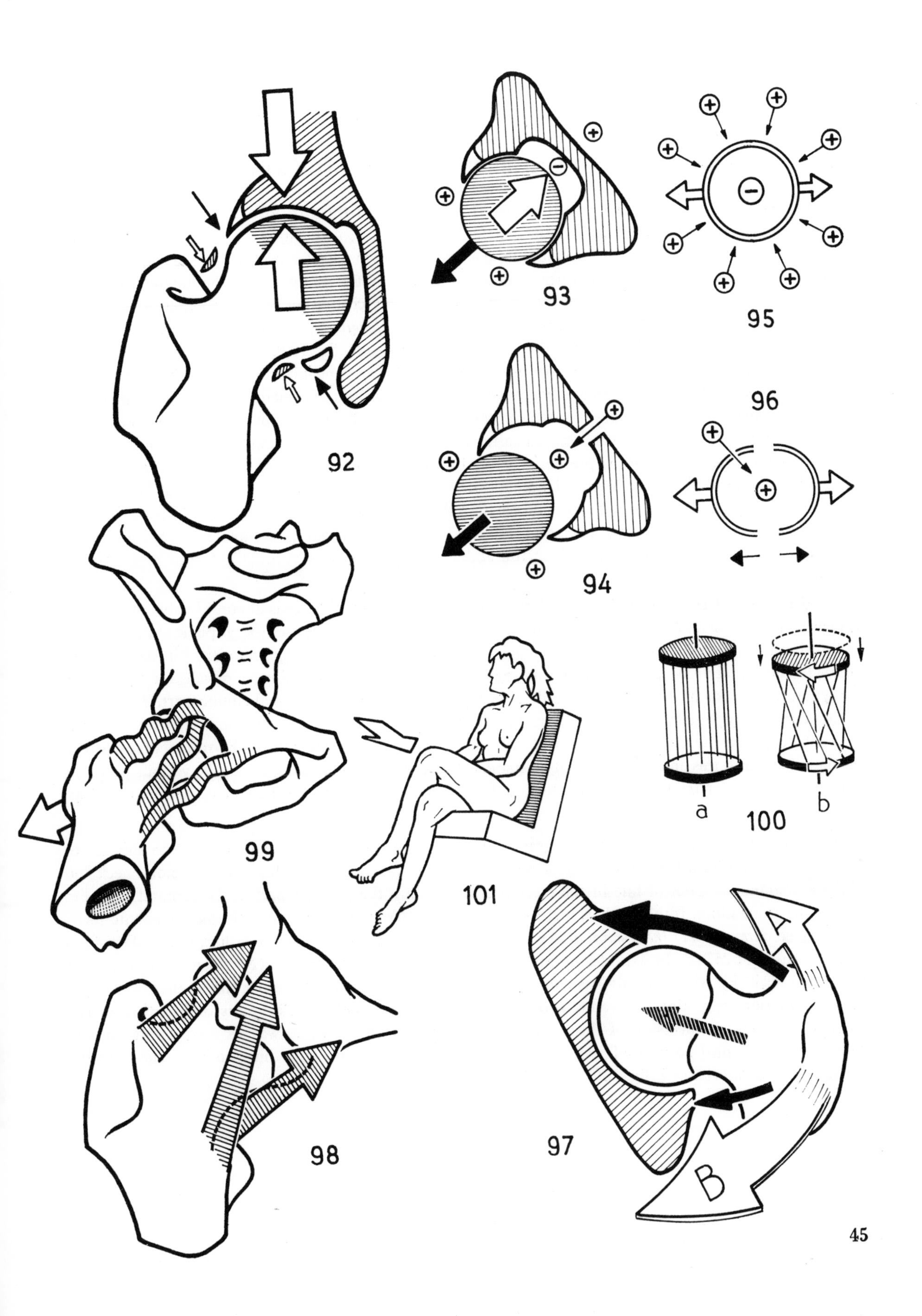

92
93
95
96
94
a
100
b
99
101
A
98
97
B

THE MUSCULAR AND BONY FACTORS INFLUENCING THE STABILITY OF THE HIP

The periarticular muscles are essential for the stability of the hip joint on condition, however, that they run transversely. In effect (fig. 102) *the muscles, running roughly parallel to the femoral neck*, keep the femoral head in contact with the acetabulum e.g. the pelvitrochanteric muscles—the piriformis (1) and the obturator externus (2) only shown here; the glutei, especially the minimus and the medius (3), which possess a powerful component of force (black arrow) producing coaptation. These muscles are therefore called the muscles of apposition of the hip. On the other hand, *longitudinal muscles*, like the adductors (4), tend to dislocate the femoral head above the acetabulum (right side, fig. 102), especially if the roof of the acetabulum is everted. This acetabular malformation is present in congenital dislocation of the hip and is easily recognised in an anteorposterior radiogram of the pelvis (fig. 103). Normally the angle of Hilgenreiner between the horizontal line running through the cartilages at level Y and the line running tangential to the acetabular roof is 25° in the neonate and 15° after the first year; when this angle exceeds 30° congenital malformation of the acetabulum is present. Dislocation is recognised by the upward displacement of the 'nucleus' of the head above the Y line (landmark of Putti) and the inversion of the angle of Wiberg (see fig. 36). In the presence of an acetabular malformation the adductors (4′) can produce dislocation, the more so when the limb is adducted (fig. 102); on the other hand, the 'dislocating' component of the adductors decreases with increased abduction until in full abduction the adductors eventually favour apposition of the surfaces (fig. 104).

The **direction of the femoral neck** in both frontal and horizontal planes is of considerable importance in maintaining the stability of the joint. It has been shown (p. 24) that in the frontal plane the axis of the neck forms an angle of 120° to 125° with the axis of the shaft (fig. 105, a: diagram of the hip seen from in front.) In congenital dislocation of the hip this angle can reach 140° (b) producing a coxa valga so that during adduction (c) the axis of the neck has already a 'head start' of 20° over its normal counterpart. Therefore a 30° adduction in a pathological hip (P) corresponds to a 50° adduction in a normal hip. Now adduction is known to enhance the dislocating action of the adductors. Hence **coxa valga promotes dislocation.** On the other hand, this abnormal hip will be stabilised in abduction; hence the use of the various positions of immobilisation for treatment of the congenital dislocation of the hip, the first being abduction at 90° (fig. 106).

In the horizontal plane (fig. 107: diagram of the hip seen from above) the angle between the axis of the femoral neck and the frontal plane has a mean value of 20° (a). Because the axes of the femoral neck and of the acetabulum are out of line in the erect posture (p. 26), the anterior part of the femoral head lies outside the acetabulum. If this angle is increased to, say, 40° (b) and the femoral neck runs more anteriorly, this is called *anteversion of the femoral neck* and the head is more liable to anterior dislocation. In fact, for a lateral rotation of 25° (c) the axis of a normal neck falls within the acetabulum (N) whereas the axis of an anteverted neck (P), which already has a 20° 'headstart', will fall on the acetabular rim so that the head is liable to anterior dislocation. Therefore **anteversion of the femoral neck favours dislocation of the joint**. Conversely, retroversion of the neck promotes stability in the same way as medial rotation (d). This is why the third position of reduction of congenital dislocation of the hip combines the 'straight' position and *medial rotation* (fig. 106).

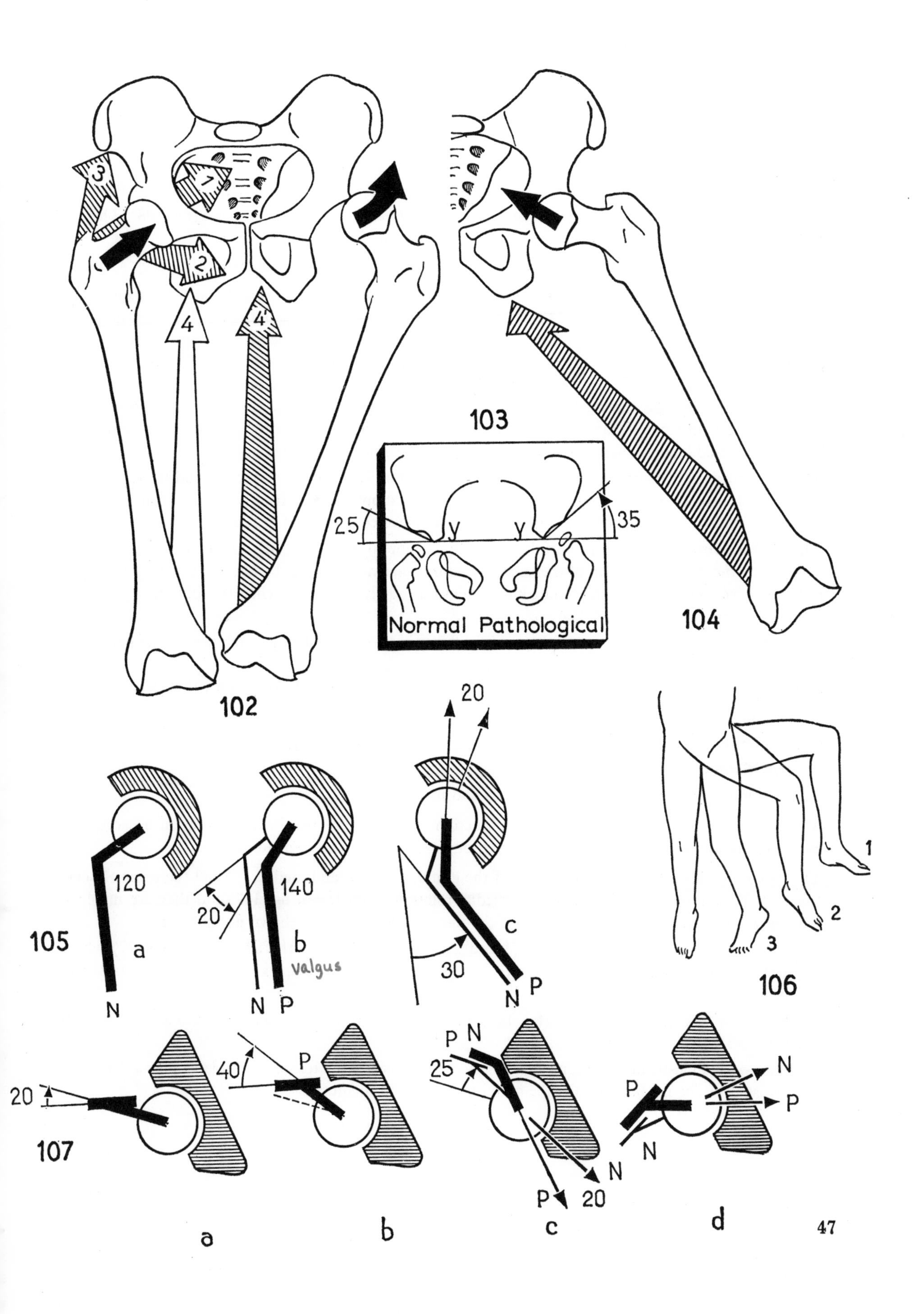
3
1
2
4
4'
103
25
35
y
y
Normal Pathological
104
102
20
120
140
20
105
a
b
valgus
30
c
N
N P
N P
1
2
3
106
20
40
P
P N
25
P
N
P
N
N
P
20
107
a
b
c
d

THE FLEXOR MUSCLES OF THE HIP

These muscles lie *anterior to the frontal plane, which passes through the centre of the joint* (fig. 108). They all run *anterior to the axis of flexion and extension* XX′ lying in this frontal plane.

There are many flexor muscles of the hip and the most important are the following (fig. 109):

The **psoas** (1) and the **iliacus** (2) share a common tendon of insertion into the lesser trochanter; this tendon bends sharply at the level of the iliopubic eminence. The iliopsoas muscle is the most powerful of the flexors and has the longest range (the highest fibres of the psoas being inserted into the twelfth thoracic vertebra). Its action as an adductor is challenged by many authors in spite of its course medial to the anteroposterior axis. This lack of any adductor action could be due to the fact that the apex of the lesser trochanter lies on the mechanical axis of the lower limb (fig. 48). However, in support of its adductor action, one notes that on the skeleton the lesser trochanter is nearest to the iliopubic eminence during flexion-adduction-lateral rotation. The iliopsoas also produces lateral rotation.

The **sartorius** (3) is mainly a flexor of the hip and secondarily produces abduction and lateral rotation (fig. 110); it also acts on the knee (flexion and medial rotation: p. 132). It is fairly powerful (muscular pull equivalent to 2 kg. weight) and nine-tenths of its power is expended in flexion.

The **rectus femoris** (4) is a powerful flexor (equivalent to 5 kg. weight) but its action on the hip depends on the degree of flexion of the knee. It is more efficient the more the knee is flexed (p. 128). It is particularly so in movements combining knee extension and hip flexion, as when the limb moves forward during walking (fig. 111).

The **tensor fasciae latae** (5) is a fairly powerful flexor in addition to being a stabiliser of the pelvis and an abductor of the hip (p. 56).

Some muscles are only *accessory flexors* of the hip but their contribution to flexion is not negligible:

the **pectineus** (6) which is primarily an adductor;

the **adductor longus** (7), primarily adductor but also partially flexor (p. 66);

the **gracilis** (8);

the most anterior fibres of the glutei, minimus and medius (9).

All these flexors of the hip can produce adduction/abduction or lateral/medial rotation as accessory movements and they can be divided into two groups according to these actions:

The *first group* includes the anterior fibres of the glutei, minimus and medius (9) and the tensor fasciae latae (5). They produce flexion-abduction-medial rotation (right thigh, fig. 109) and they are involved alone or predominantly in the production of the footballer's movement shown in figure 112.

The *second group* includes the iliopsoas (1 and 2), pectineus (6) and the adductor longus (7) which produce flexion-adduction-lateral rotation (left thigh, fig. 109); this complex movement is illustrated in figure 113.

During simple flexion, as in walking (fig. 111), these two groups must act as a *balanced set of synergists and antagonists.*

In flexion-adduction-medial rotation (fig. 114) the adductors and the tensor fasciae latae play a dominant role, assisted by the medial rotators, the glutei minimus and medius.

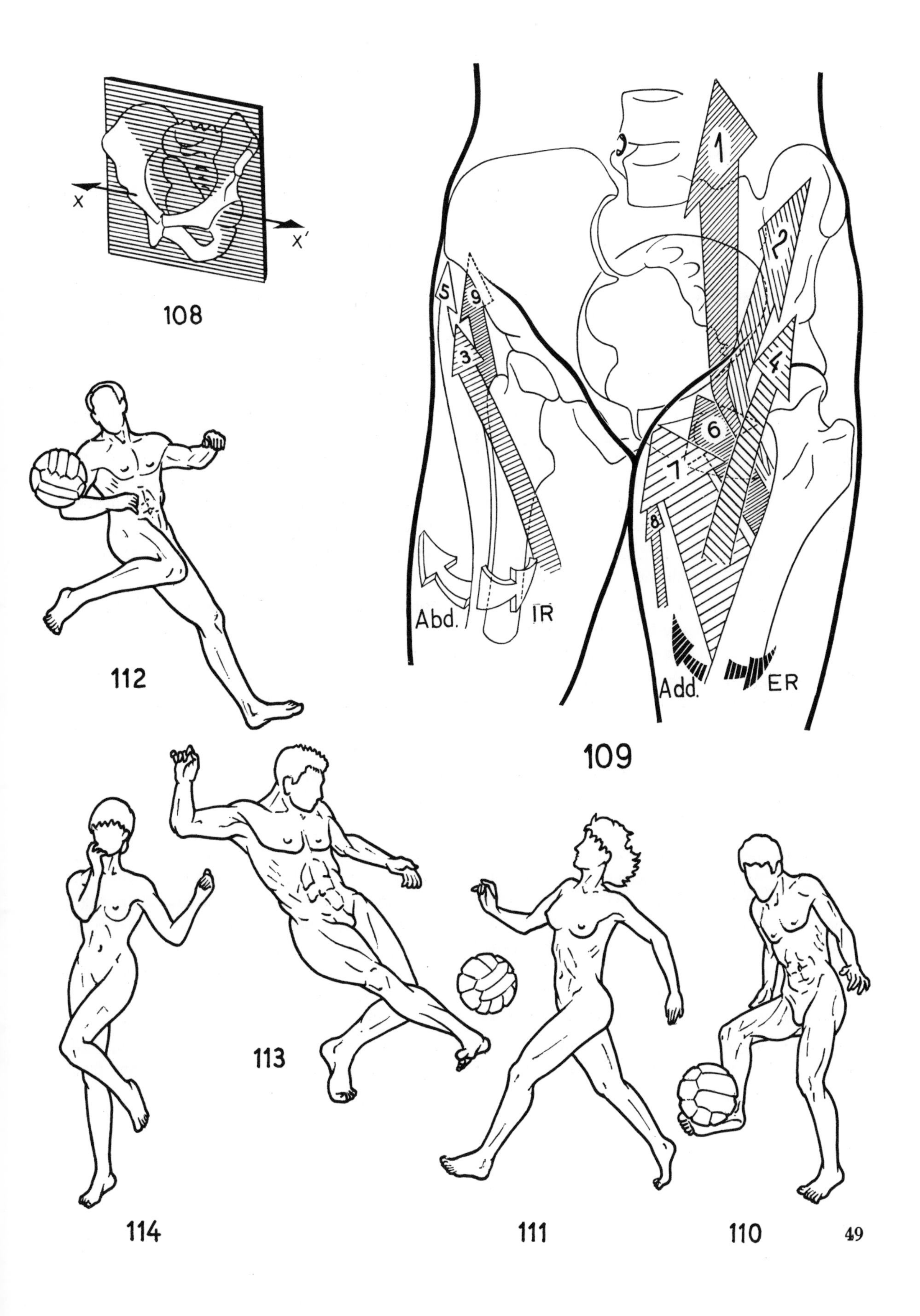
X
X'
108
1
2
3
4
5
6
7
8
9
Abd.
IR
Add.
ER
109
112
113
114
111
110

THE EXTENSOR MUSCLES OF THE HIP

These muscles *lie behind the frontal plane that passes through the centre of the joint* (fig. 115) and contains the transverse axis XX′ of flexion and extension.

There are **two main groups of muscles**: the one group is inserted into the femur and the other in the vicinity of the knee joint (fig. 116).

Of the *first group* the **gluteus maximus** (1 and 1′) is the most important. It is the most powerful muscle of the body (force equivalent to 34 kg. weight; contraction length of muscle = 15 cm.), and is also the biggest (66 cm.[2] in cross-section) and, naturally, the strongest (its static power is equivalent to 238 kg. weight). It is assisted by the *most posterior fibres of the glutei, medius* (2) *and minimus* (3). These muscles are also lateral rotators (p. 62).

The *second group* consists essentially of the **hamstring muscles** i.e. biceps femoris (4), semitendinosus (5), seminembranosus (6); their power is equivalent to 22 kg. weight, i.e. two-thirds that of the gluteus maximus. They are biarticular muscles and *their efficiency at the hip depends on the position of the knee*: the locking of the knee in extension enhances their extensor action at the hip; this suggests a synergism between the hamstrings and the quadriceps femoris (especially the rectus). This group also includes some adductors, especially the adductor magnus (7) which is an *accessory* extensor of the hip (see page 60).

The extensor muscles of the hip have *secondary actions* depending on their position relative to the antero-posterior axis YY′ of adduction and abduction:

those running *superior to the axis YY′* produce abduction along with the extension as in the dancing movement shown in figure 117. These muscles include the most posterior fibres of the glutei, minimus (3) and medius (2) and the most superior fibres of the gluteus maximus (1′);

those running *inferior to the axis YY′* produce adduction and extension as in the movement shown in fig. 118. These muscles are the hamstrings, the adductors (i.e. those lying behind the frontal plane) and the bulk of the gluteus maximus (1).

To produce pure extension (fig. 119) i.e. without associated adduction and abduction, these two muscle groups are thrown into balanced contraction as synergists and antagonists.

The extensors of the hip joint play an essential part in *stabilising the pelvis in the anteroposterior direction* (fig. 120):

When the pelvis is tilted posteriorly (a), i.e. in the direction of extension, it is stabilised only by the tension of the iliofemoral ligament (IFL) which limits extension (p. 36); there is a position (b) where the centre of gravity (c) of the pelvis lies directly above the centre of the hip. The flexors and extensors are not active but the equilibrium is unstable; when the pelvis is tilted anteriorly (c) the centre of gravity (c) comes to lie in front of the transverse axis of the hips and the hamstrings (H) are the first to contract so as to straighten the pelvis; when the pelvis is tilted very far anteriorly (d) the gluteus (G) maximus contracts powerfully as well as the hamstrings, which are more efficient the greater the degree of knee extension (standing with trunk bent forwards and the hands touching the feet).

During normal walking extension is produced by the hamstrings and *the gluteus maximus is not involved*. However, when one is running, jumping or walking up a slope, the gluteus maximus is essential and plays an important part.

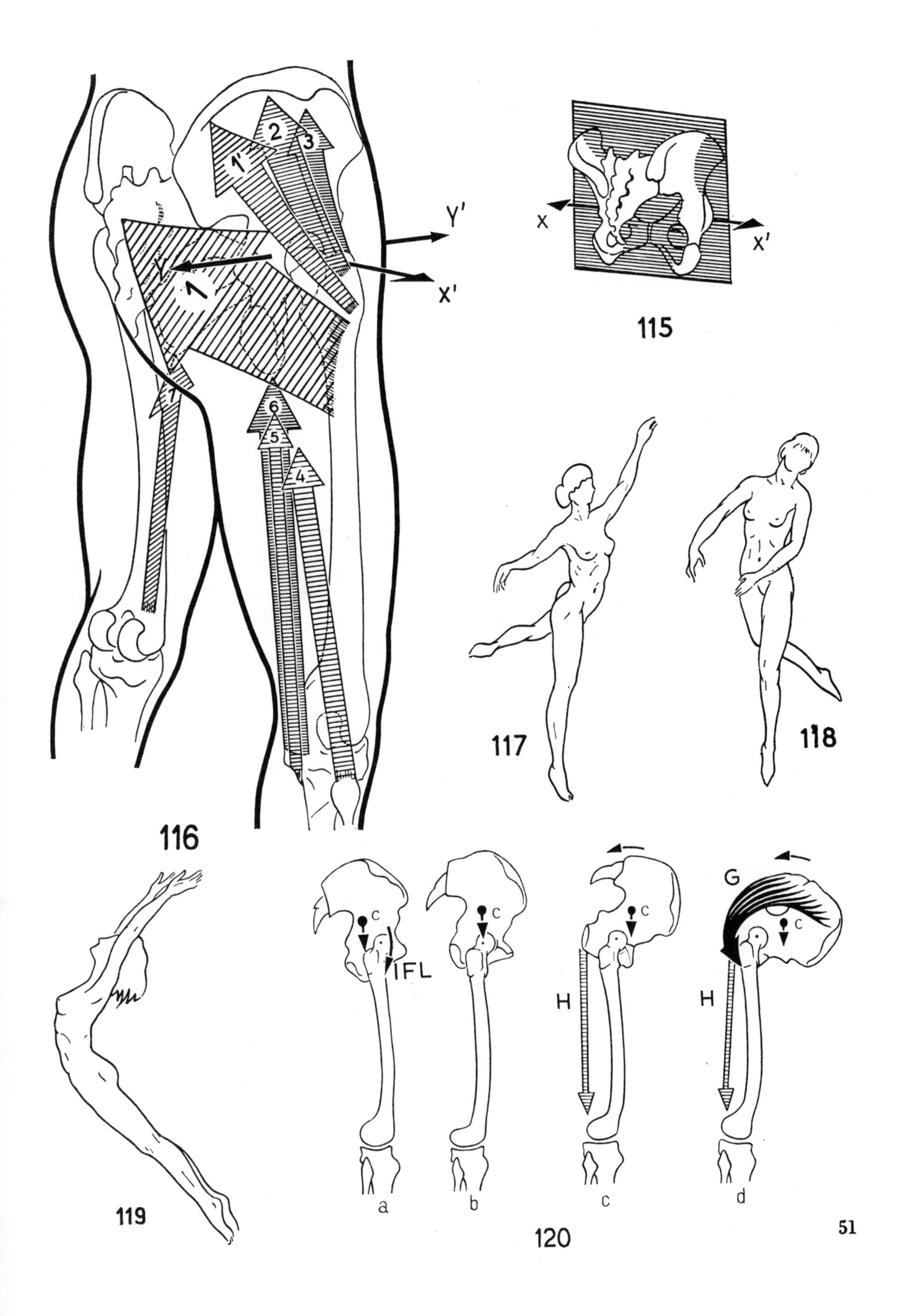

115

116

117

118

119

120

ABDUCTOR MUSCLES AND THE TRANSVERSE STABILITY OF THE PELVIS

These muscles **generally lie lateral to the sagittal plane which traverses the centre of the joint** (fig. 121) and run laterally and superiorly to the **anteroposterior axis YY′ of adduction and abduction** contained in that sagittal plane.

The main abductor muscle is the **gluteus medius** (1). It has a cross-sectional area of 40 $cm.^2$ and it contracts by 11 cm. It can produce a force equivalent to 16 kg. weight. It is highly efficient because it is almost perpendicular to its lever arm OT (fig. 122). It is also essential, along with the gluteus minimus, for the stabilisation of the pelvis in the transverse direction (p. 56).

The **gluteus minimus** (2) is essentially an abductor (fig. 123).

It has a cross-sectional area of 15 $cm.^2$ and it shortens by 9 cm. during contraction. It can produce a force equivalent to 5 kg. weight, i.e. about one-third that of the gluteus medius.

The **tensor fasciae latae** (3) is a powerful abductor of the hip when it is in the erect position. Its muscular power is about half of that of the gluteus medius (7·6 kg. weight) but its lever arm is much longer than that of the gluteus medius. It also acts to stabilise the pelvis.

The **gluteus maximus** (4) produces abduction only with its highest fibres (the bulk of the muscle produces adduction) and its superficial fibres, which form part of the so-called 'deltoid of the hip' (fig. 127).

The **piriformis** (5) is undoubtedly an abductor but its action cannot be easily demonstrated experimentally because of its deep location.

According to their secondary movements of flexion/extension and adduction/abduction, these abductor muscles can be classified into two groups:

The *first group* includes all the muscles lying anterior to the frontal plane running through the centre of the joint, i.e. tensor fasciae latae, the anterior fibres of the gluteus medius, and the bulk of the gluteus minimus. These muscles, whether they contract alone or are assisted by weaker fellows, produce **abduction-flexion-medial rotation** (fig. 124).

The *second group* consists of the posterior fibres of the glutei, minimus and medius (the fibres lying posterior to the frontal plane) and the abductor fibres of gluteus maximus. These muscles, whether they contract alone or are assisted by weaker fellows, produce **abduction-extension-lateral rotation** (fig. 125).

To obtain **pure abduction** (fig. 126), i.e. without any other associated movements, these two groups of muscles must be activated as a balanced couple of synergists-antagonists.

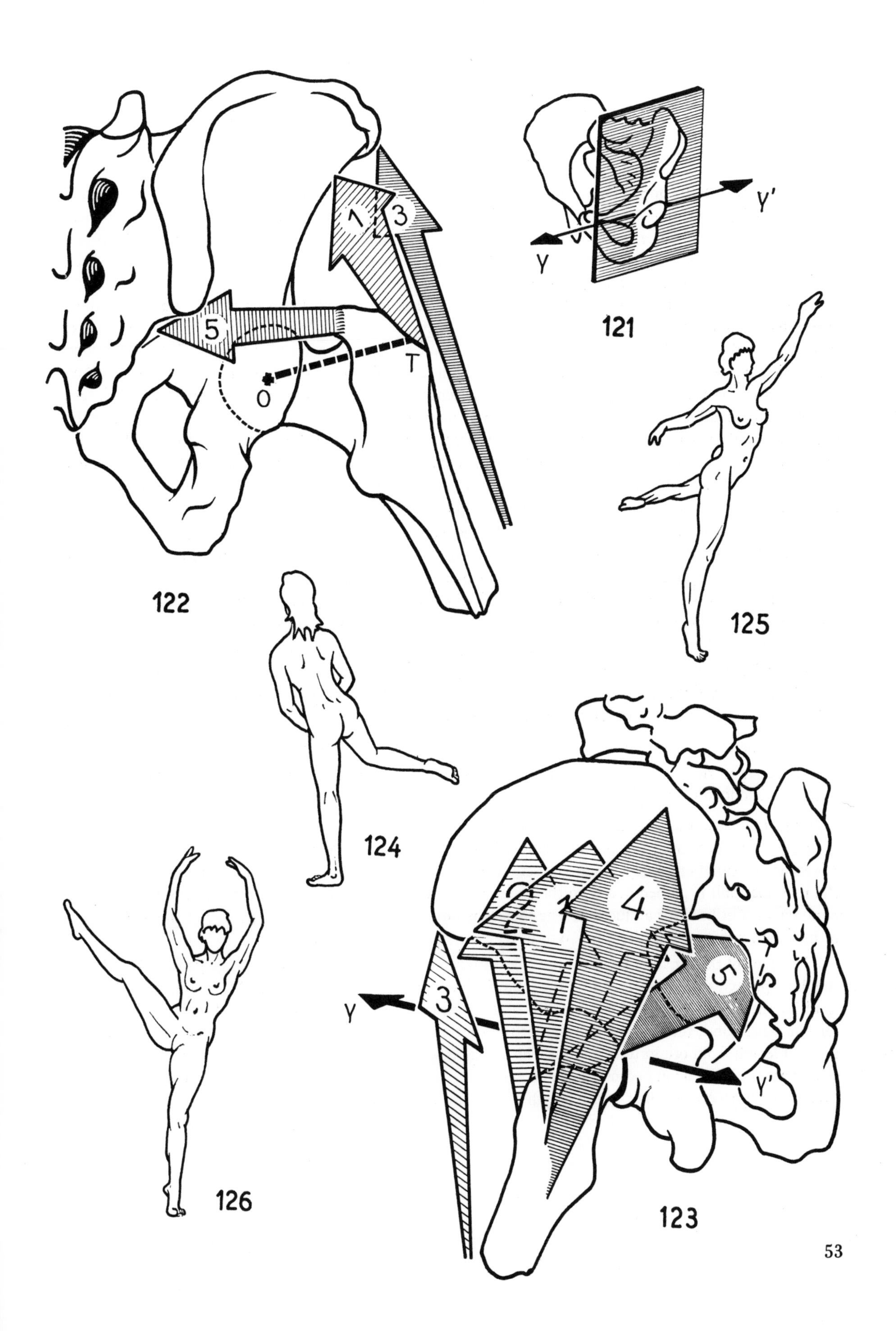

122
121
125
124
126
123

The **'deltoid of the hip'** (Farabeuf) consists of a wide muscular fan (fig. 127) covering the external aspect of the hip joint. It owes its name to its *triangular shape* with apex pointing inferiorly and its anatomical and functional resemblance to the deltoid muscle of the shoulder joint. However it is not made up of a continuous sheet of muscle but of two muscle bellies lying at the anterior and posterior edges of the triangle. Anteriorly, the *tensor fasciae latae* (1), arising from the anterosuperior iliac spine (2), runs obliquely inferiorly and posteriorly. Posteriorly, the *superficial fibres of the gluteus maximus* (3), arising from the posterior third of the iliac crest and the dorsum of the sacrum and coccyx, run obliquely inferiorly and anteriorly. These two muscles are inserted respectively into the anterior and posterior borders of the iliotibial tract (4) which splits into superficial and deep layers to invest those muscles; the iliotibial tract is a long strong band which forms part of the fascia lata. After insertion of the tensor fasciae latae and the superior fibres of gluteus maximus, the iliotibial tract becomes the tendon of insertion of the 'deltoid' (5) which is attached to the lateral aspect of the lateral tibial condyle on the tubercle of Gerdy (6). Between the tensor fasciae latae and the gluteus maximus the deep fascia of the thigh (7) covers the gluteus medius. Naturally the two muscular parts of the 'deltoid' can contract separately but, when they contract in a balanced fashion, the tendon is pulled along its long axis and the deltoid produces pure abduction.

The efficiency of the gluteus minimus and gluteus medius is *conditioned by the length of the femoral neck* (fig. 128). If the head were placed directly on the shaft, the total range of abduction would be considerably increased but the lever arm OT of the gluteus medius would be reduced by one-third as would its power. This therefore gives a natural 'explanation' why the femoral head overhangs the shaft: this mechanical arrangement is weaker and decreases the range of abduction but it ensures the greater efficiency of the gluteus medius which is essential for stabilising the pelvis in the transverse direction.

The action of gluteus medius (fig. 129) on the lever arm of the femoral neck varies with the degree of abduction. When the hip is straight (a) the muscular pull F does not act at right angles to the lever arm OT, and so can be resolved into two vector components:

1. f'' acting towards the centre of the joint (i.e. centripetal) and promoting coaptation (fig. 102);
2. f' acting at right angles (i.e. tangential) and so representing the effective force of the muscle at the start of abduction.

Subsequently, as abduction gains range (b), the vector f'' tends to decrease as f' increases. Therefore the gluteus medius becomes progressively more efficient in producing abduction than in securing coaptation. It reaches maximal efficiency, when abduction has approximately a 35° range, i.e. when the direction of the muscular pull is perpendicular to the lever arm OT_2 and so f' coincides with F and the full energy of contraction is utilised for abduction. The muscle is now shorter by a length of T_1 T_2, i.e. about one-third of its length but it still has left one-sixth of its excursion during contraction.

The action of the **tensor fasciae latae** (fig. 130) can be studied in the same way (a). Its force F acting on the iliac spine C_1 can be resolved into the following two vectors: f_1'' centripetal and f_2' tangential and active in tilting the pelvis. As abduction increases (b), f'_2 increases but never equals the total force F of the muscle. On the other hand, the diagram shows that the shortening of the muscle C'_1 C_2 is only a tiny fraction of its total length from origin to insertion. This explains why the muscular portion of the muscle is so short as compared to its tendon because it is a known fact that the maximal excursion of a muscle during contraction does not exceed one half of the length of its contractile fibres.

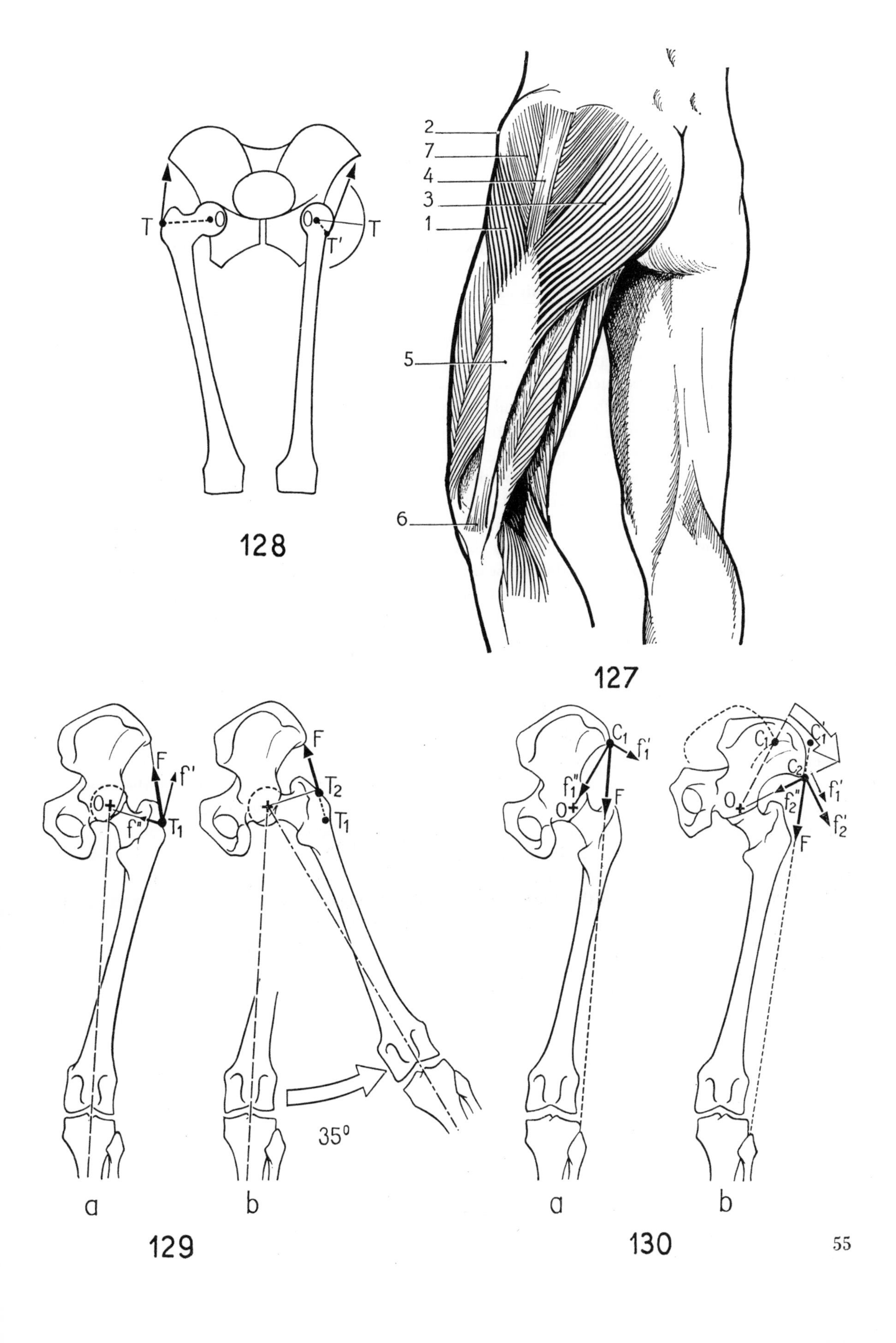
2
7
4
3
1
5
6
127
T
O
T
T'
128
F
f'
O
f''
T1
a
F
T2
T1
b
35°
129
C1
f'1
f''1
O
F
a
C1
C'1
C2
O
f''2
f'1
f'2
F
b
130

THE STABILISATION OF THE PELVIS IN THE TRANSVERSE DIRECTION

When the **pelvis is supported on both sides** (fig. 131), its stability in the transverse direction is secured by the simultaneous contraction of the ipsilateral and contralateral adductors and abductors. When these antagonistic actions are properly balanced (a) the pelvis is stabilised in the position of symmetry, as in the military position of standing to attention. If the abductors predominate on one side and the adductors on the other (b), the pelvis is tilted laterally towards the side of adductor predominance. If muscular equilibrium cannot be restored at this point the subject will fall to that side.

When the **pelvis is supported only one limb** (fig. 132), the stability of the pelvis is provided solely by the action of the ipsilateral abductors, since the weight of the body P, acting through the centre of gravity, will tend to tilt the pelvis at the supporting hip. The pelvis can therefore be compared to a lever system of type I (fig. 133), where the fulcrum is the supporting hip O, the disturbing force is the weight of the body P acting through the centre of gravity G and the restoring force is the muscular pull of gluteus medius (GMe) acting at the external iliac fossa E. To keep the pelvis horizontal when supported on one leg, the muscular force of gluteus medius must cancel the force exerted by the body weight, taking into account that the lever arms OE and OG are not equal in length. In this action the gluteus medius is powerfully assisted by gluteus minimus and the tensor fasciae latae (fig. 132).

If there is insufficiency of any one of these muscles (fig. 132, b) the body weight acting at G is not properly counterbalanced and the pelvis tilts to the opposite side to form an angle α with the horizontal, which is directly proportional to the severity of the muscular insufficiency. The tensor fasciae latae stabilises not only the pelvis but also the knee: therefore (p. 110) it is effectively an **active** collateral ligament so that its paralysis in the long run leads to an abnormal widening of the knee joint interspace laterally (angle β).

Stabilisation of the pelvis by the glutei and the tensor fasciae latae is essential for normal walking (fig. 134). While the pelvis is supported on one limb, the transverse axis of the pelvis, represented by the interiliac line, stays horizontal and parallel to the line joining the two shoulders. If these muscles are paralysed on the side supporting the pelvis (fig. 135), the pelvis is tilted to the opposite side. This would result in a fall to that side, if the whole trunk were not bent towards the supporting side and the line of the shoulders tilted the same way. This combination of movements during walking—i.e. tilting of the pelvis towards the unsupported side and bending of the upper trunk towards the supported side—is very characteristic and is used clinically (sign of Duchenne-Trendelenburg) to demonstrate complete or partial paralysis of the glutei, minimus and medius.

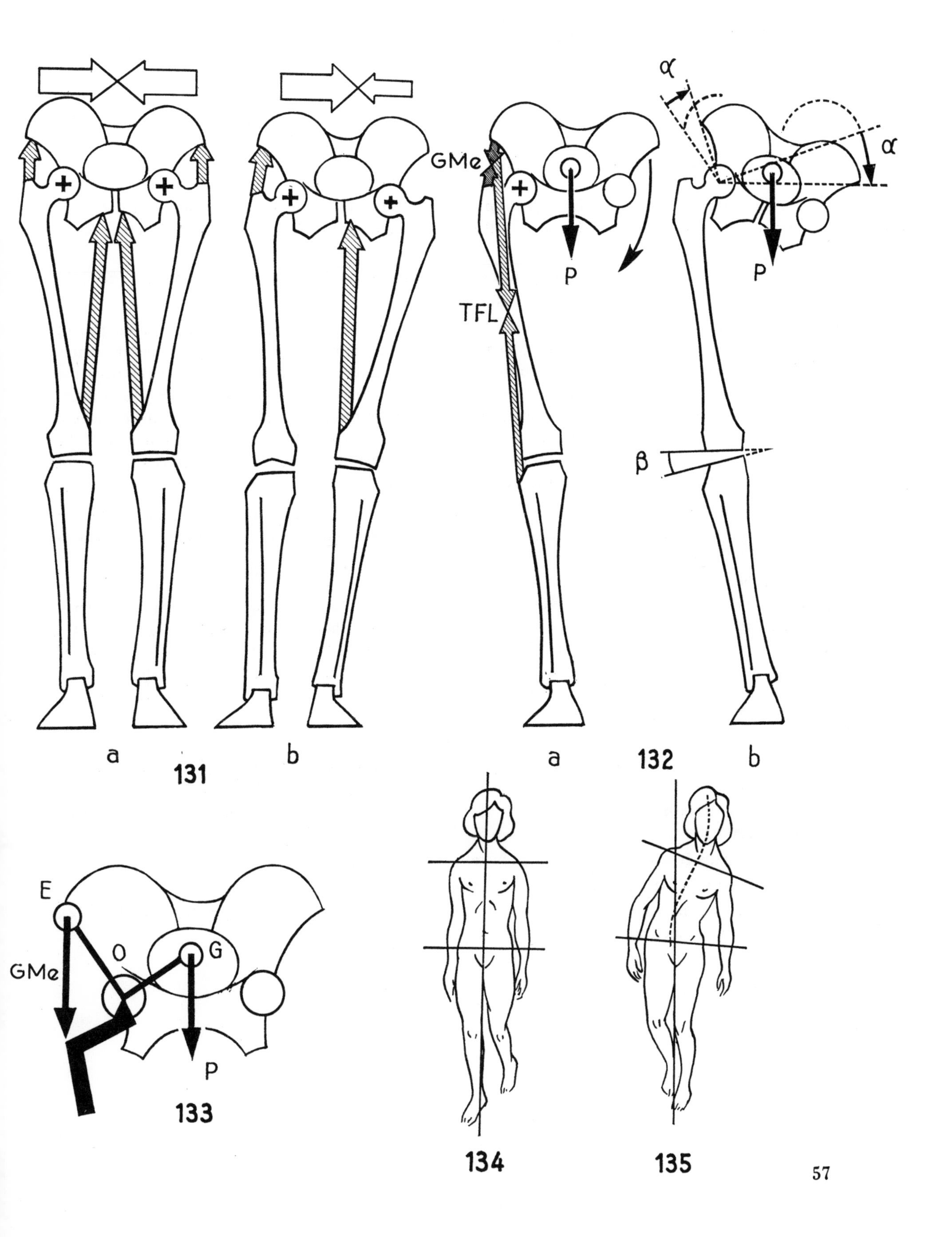
GMe
TFL
P
P
α
α
β
a
b
131
a
132
b
E
O
G
GMe
P
133
134
135

ADDUCTOR MUSCLES OF THE HIP

These **lie generally medial to the sagittal plane, which traverses the centre of the joint** (fig. 136). In any case they run inferior and medial to the anteroposterior axis YY′ of adduction-abduction which lies in the sagittal plane.

The adductor muscles are particularly **numerous** and **powerful**. When seen from behind (fig. 137) they form a large fan covering the whole length of the femur:

the **adductor magnus** (1) is the most powerful (equivalent to 13 kg. weight). Its peculiar arrangement (fig. 138) is due to the fact that its most medial fibres arising from the ramus of the pubis and the ramus of the ischium are inserted the most proximally on the femur while its most lateral fibres from the ischial tuberosity are inserted the most distally on the linea aspera. Therefore its superior (2) and intermediate (1) fibres form a sort of gutter, concave posterolaterally; this can be seen in the diagram because the superior fibres are considered transparent and the hip has been disarticulated with lateral rotation of the femur. In this gutter (inset representing a section taken at the level of the arrow) runs a third set of fibres (the inferior fibres), which constitute a distinct muscle belly, often called the 'third adductor' (3).

Owing to this arrangement of fibres the relative passive lenghtening of the muscle fibres during abduction is reduced so that it allows a greater degree of abduction while retaining the efficiency of the muscle. This is illustrated in figure 139 as follows:

side A shows the real arrangement of the fibres;

side B shows the real arrangement of the fibres (broken line) and the 'simplified' arrangement (dotted line): the innermost fibres are shown to be the lowest at insertion and the outermost fibres the highest (i.e. the exact opposite of the arrangement in life). These two arrangements (i.e. real and 'simplified' are illustrated in adduction (Add) and abduction (Abd). The lengthening of the fibres during adduction-abduction in the case of the real arrangement (black strip) and of the 'inverse' or 'simplified' arrangement (white strip) is clearly different.

The **gracilis** (4), forming the internal border of the muscular fan.

The **semimembranosus** (5), the **semitendinosus** (6), and **biceps femoris** (7), though primarily extensors of the hip and flexors of the knee, also play an important role in adduction;

the **gluteus maximus** (8): the large bulk of its fibres (including all those lying above the axis YY′) are adductor;

the **quadratus femoris** (9) produces adduction and external rotation;

the **pectineus** (10) has the same action as the quadratus;

the **obturator internus** (11), assisted by the gemelli (not shown), and the **obturator externus** (12), are secondary adductors.

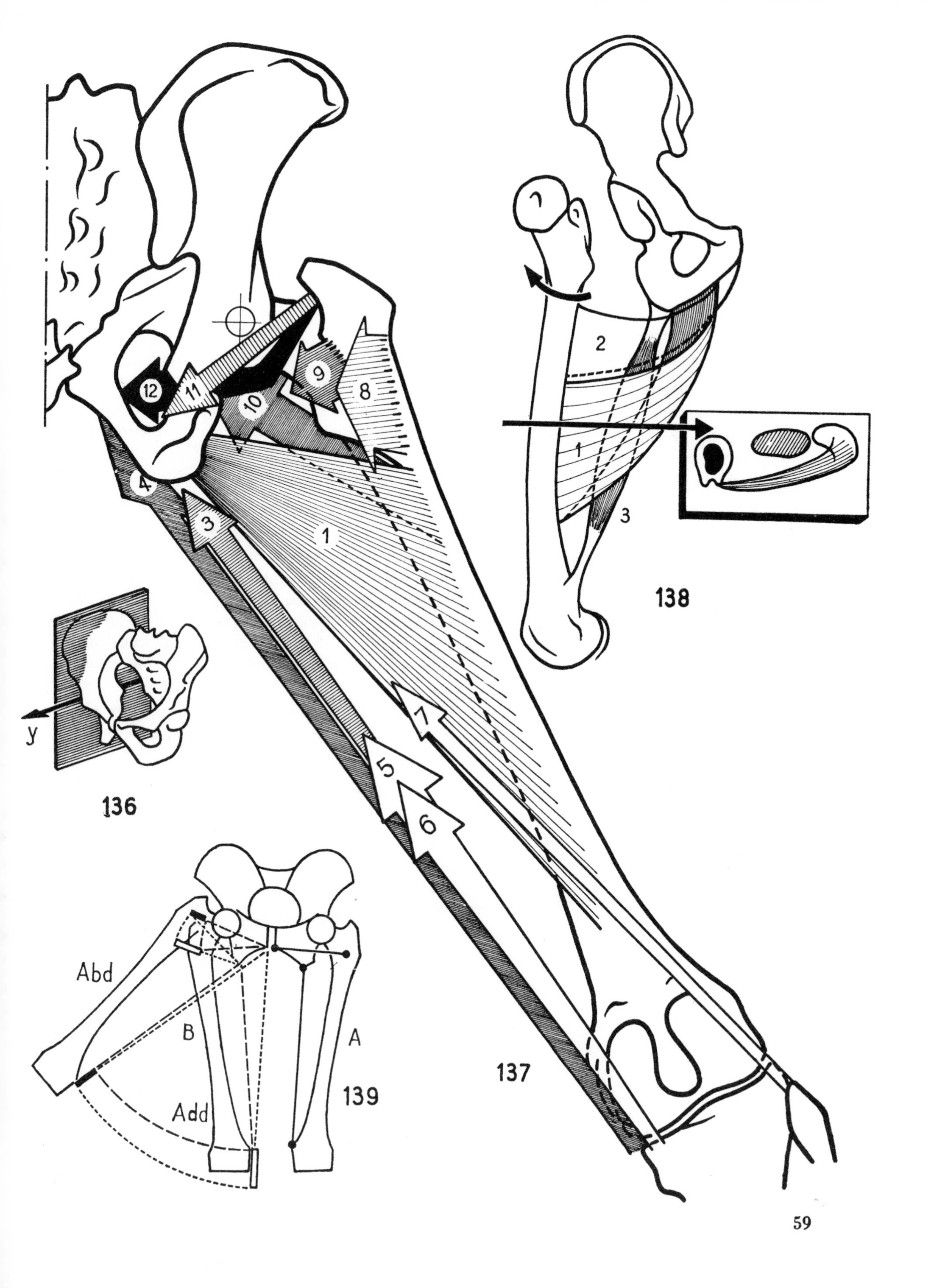
12
11
10
9
8
4
3
1
7
5
6
2
1
3
138
y
136
Abd
B
A
Add
139
137

The adductor muscles are seen from the front in figure 140:

the **adductor longus** (13) whose muscular power (equivalent to 5 kg. weight) falls short of half that of the adductor magnus;

The **adductor brevis** (14): its two bundles are covered inferiorly by the adductor longus and superiorly by the pectineus;

the **gracilis** (4) which forms the internal border of the adductor compartment.

In addition to their primary adductor action the muscles also produce some measure of flexion-extension and axial rotation.

Their role in flexion and extension (fig. 141: seen from inside) depends upon the site of their origin. If they arise from the hip bone *posterior* to the frontal plane which runs through the centre of the joint (line of alternate dashes and dots), they produce extension, especially the inferior fibres of the adductor magnus (i.e. the 'third adductor') and, of course, the hamstrings. If the muscles arise from the hip bone *anterior* to that frontal plane, they produce flexion, e.g. pectineus, adductor brevis and adductor longus, upper fibres of the adductor magnus and gracilis. Note, however, that their role in flexion and extension depends also on the initial position of the hip (p. 66).

The adductors, as shown previously, are essential for the stabilisation of the pelvis when supported on both limbs; they therefore play an essential part in certain postures or movements in sport e.g. skiing (fig. 142) and horse-riding (fig. 143).

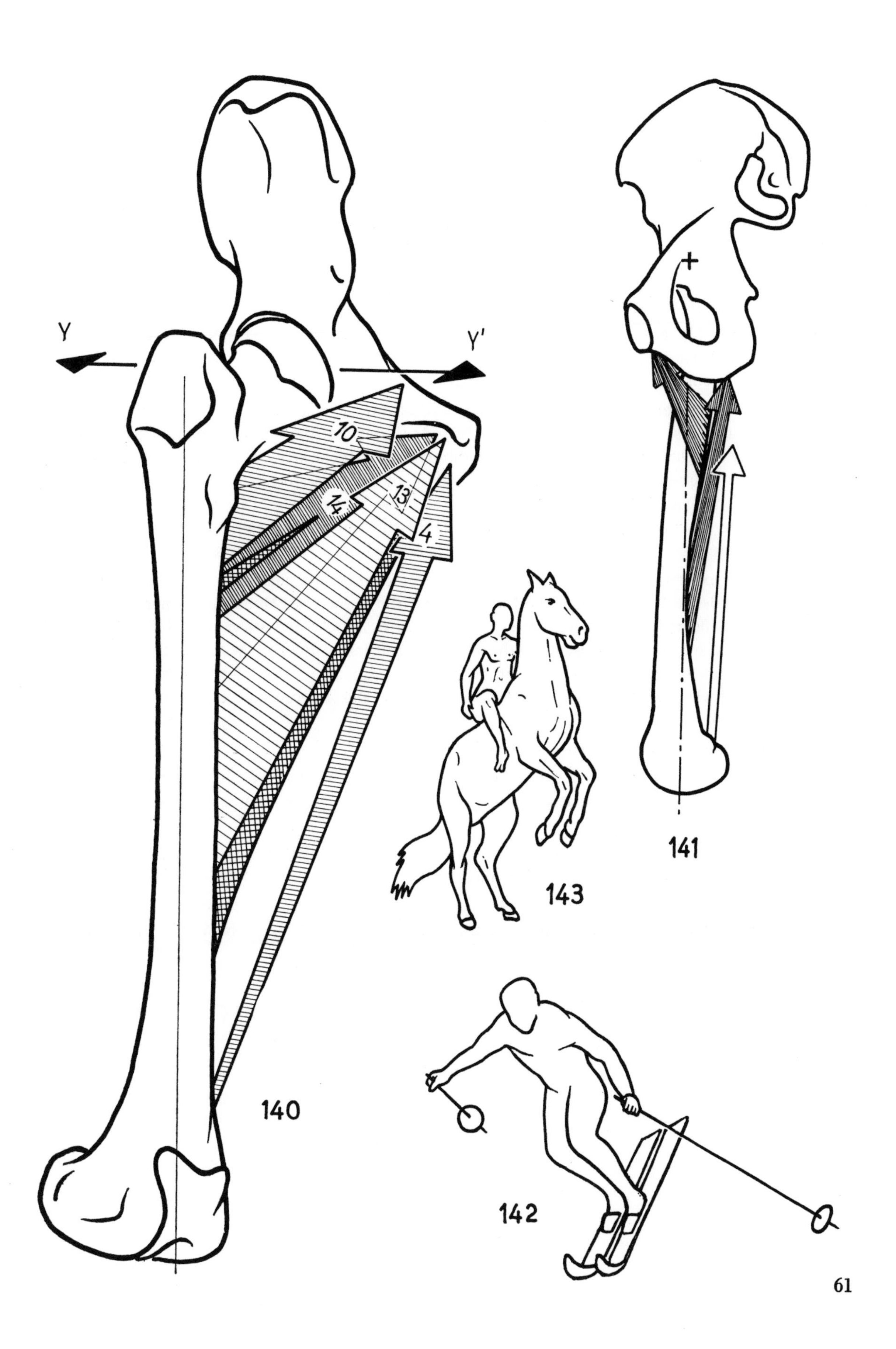
Y
Y'
10
14
13
4
140
141
142
143

THE LATERAL ROTATOR MUSCLES OF THE HIP

These are **numerous** and **powerful.** *During their course they cross the vertical axis of the hip posteriorly*, as is shown by a *horizontal section* of the pelvis passing slightly above the centre of the joint (fig. 144: seen from above). This diagram shows all the lateral rotators, which are:

the **pelvitrochanteric muscles** with lateral rotation as primary function:

the **piriformis** (1), arising from the inner surface of the sacrum, runs anteriorly and externally, emerges through the greater sciatic foramen (fig. 145: seen from behind and above) and is inserted by tendon into the superior margin of the greater trochanter;

the **obturator internus** (2) arises from the side wall of the pelvis around the obturator foramen and runs part of its course within the pelvic cavity (2′). At the level of the lesser sciatic foramen its tendon bends sharply at right angles (fig. 145), emerges through the lesser sciatic foramen and runs parallel to the piriformis to its insertion into the medial surface of the greater trochanter. Outside the pelvis the muscle is accompanied by the **two gemelli,** which are two tiny muscles arising respectively from the margins of the lesser sciatic notch. They run along the superior and inferior borders of the obturator internus and are inserted via its tendon into the greater trochanter. They have a similar action;

the **obturator externus** (3) arises from the external surface of the margins of the obturator foramen and its tendon winds posteriorly below the hip joint and runs upwards behind the femoral neck to its insertion into the floor of the trochanteric fossa. On the whole the muscle winds its way round the femoral neck and can only be seen in its entirety when the pelvis is considerably tilted on the femur (fig. 146: postero-infero-lateral view of the pelvis with the hip flexed). This explains its two main actions: it is especially a lateral rotator when the hip is flexed (p. 64) and is slightly flexor because of its winding course round the femoral neck.

Some adductor muscles are also lateral rotators:

the **quadratus femoris** (4) arising from the ischial tuberosity and inserted into the trochanteric crest and quadrate tubercle (fig. 145). It can also extend or flex the hip according to the position of the latter (fig. 153);

the **pectineus** (6) arising from the pectineal border and surface of the pubis and inserted into the femur along the line joining the lesser trochanter and the linea aspera (fig. 146). It produces adduction, flexion and lateral rotation;

the most posterior fibres of the adductor magnus also produce lateral rotation like the hamstrings (fig. 147);

the **glutei**:

the **gluteus maximus** as a whole is a lateral rotator, including its superficial (7) and deep (7′) fibres; the posterior fibres of the gluteus minimus and especially of the **gluteus medius** (8) (figs. 144 and 145).

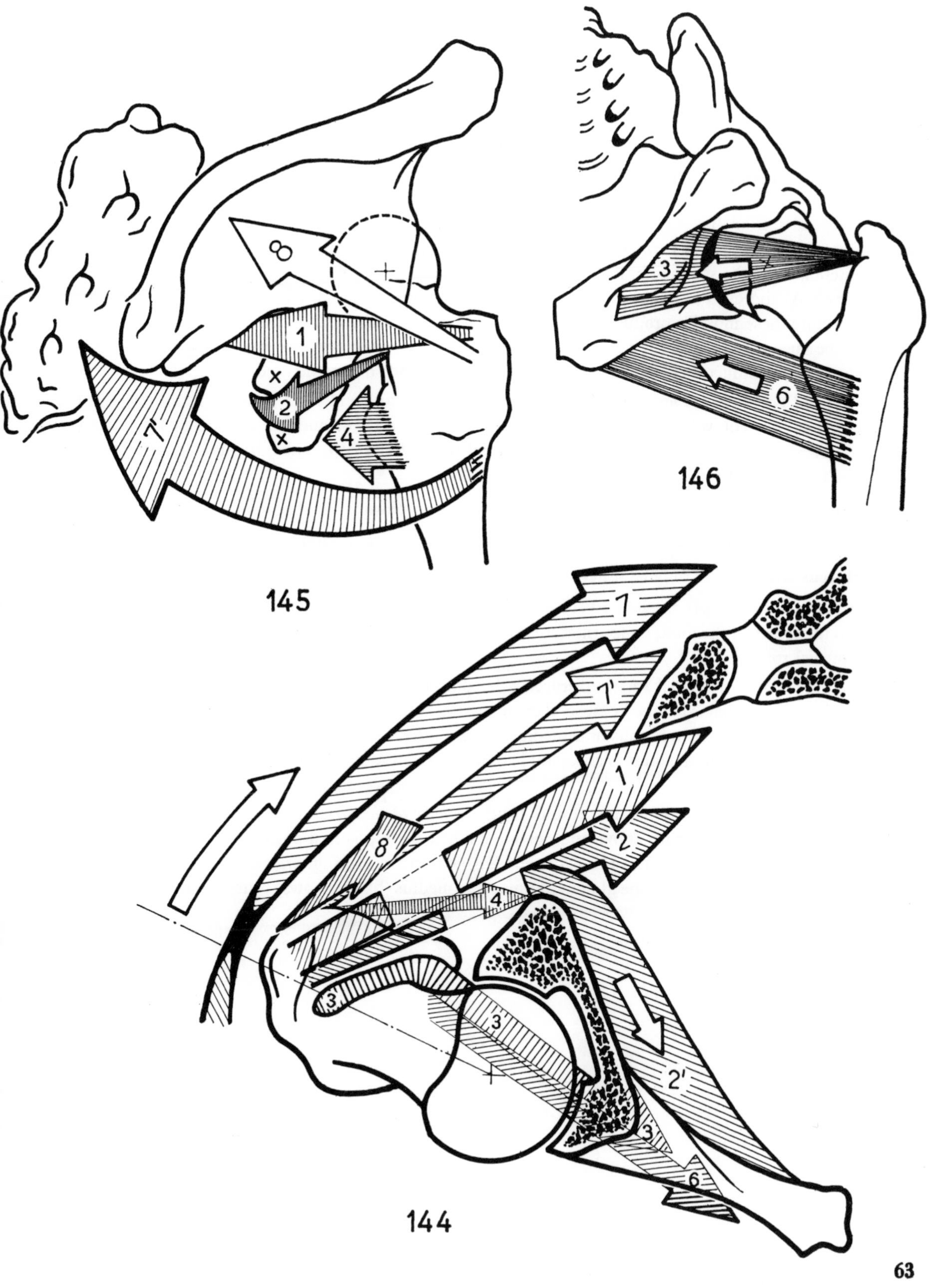
145
146
144
1
2
3
4
6
7
7'
8
2'

ROTATOR MUSCLES OF THE HIP

The horizontal section (fig. 147), passing just below the femoral head (dotted ring), shows *the rotational component of force of the hamstrings and adductors*. The horizontal projections of the *biceps femoris* (B), the semitendinosus, the seminembranosus, the '*third adductor*' (i.e. the lowest fibres of adductor magnus) (white arrow A) and even of the *adductor longus* (AL) and adductor brevis all run posterior to the vertical axis. Therefore these muscles produce lateral rotation when the lower limb turns on its long axis (fig. 48), i.e. with the knee extended and the hip and foot acting as a swivel. Note also that during medial rotation (MR) some of the adductors run anterior to the vertical axis and so become medial rotators.

The **medial rotators** are less numerous than the lateral rotators and their power is about one-third of the lateral rotators (equivalent to 54 kg. weight for the medial rotators and 146 kg. weight for the lateral rotators). These muscles run **anterior to the vertical axis of the hip**. The horizontal section (fig. 148) shows the three medial rotators of the hip:

the **tensor fasciae latae** (1) which runs superior to the anterosuperior iliac spine (S);

the **gluteus minimus** (2): practically all its fibres produce lateral rotation;

the **gluteus medius** (3): only its anterior fibres produce lateral rotation.

When medial rotation has reached 30° to 40° (fig. 149) the *obturator externus* (4) and the *pectineus* run inferior to the centre of the joint: hence they are no longer lateral rotators. The glutei, minimus and medius, are still medial rotators.

On the other hand, if medial rotation proceeds beyond 40° (fig. 150), the obturator externus and the pectineus become medial rotators because they now run anterior to the vertical axis while the tensor fasciae latae and the glutei, minimus and medius, *become lateral rotators*. This only applies when medial rotation reaches its maximum and it is an example of inversion of muscular action according to the position of the hip.

This **inversion of muscular action** results from a *change of the direction of the muscle fibres* as shown in figure 151 (seen from in front, above and outside). When the hip is in a position of exaggerated medial rotation, the obturator externus and the pectineus (striped arrows) run anterior to the vertical axis (line of alternate dashes and dots) while the glutei, minimus and medius (black arrows) course obliquely superiorly and posteriorly.

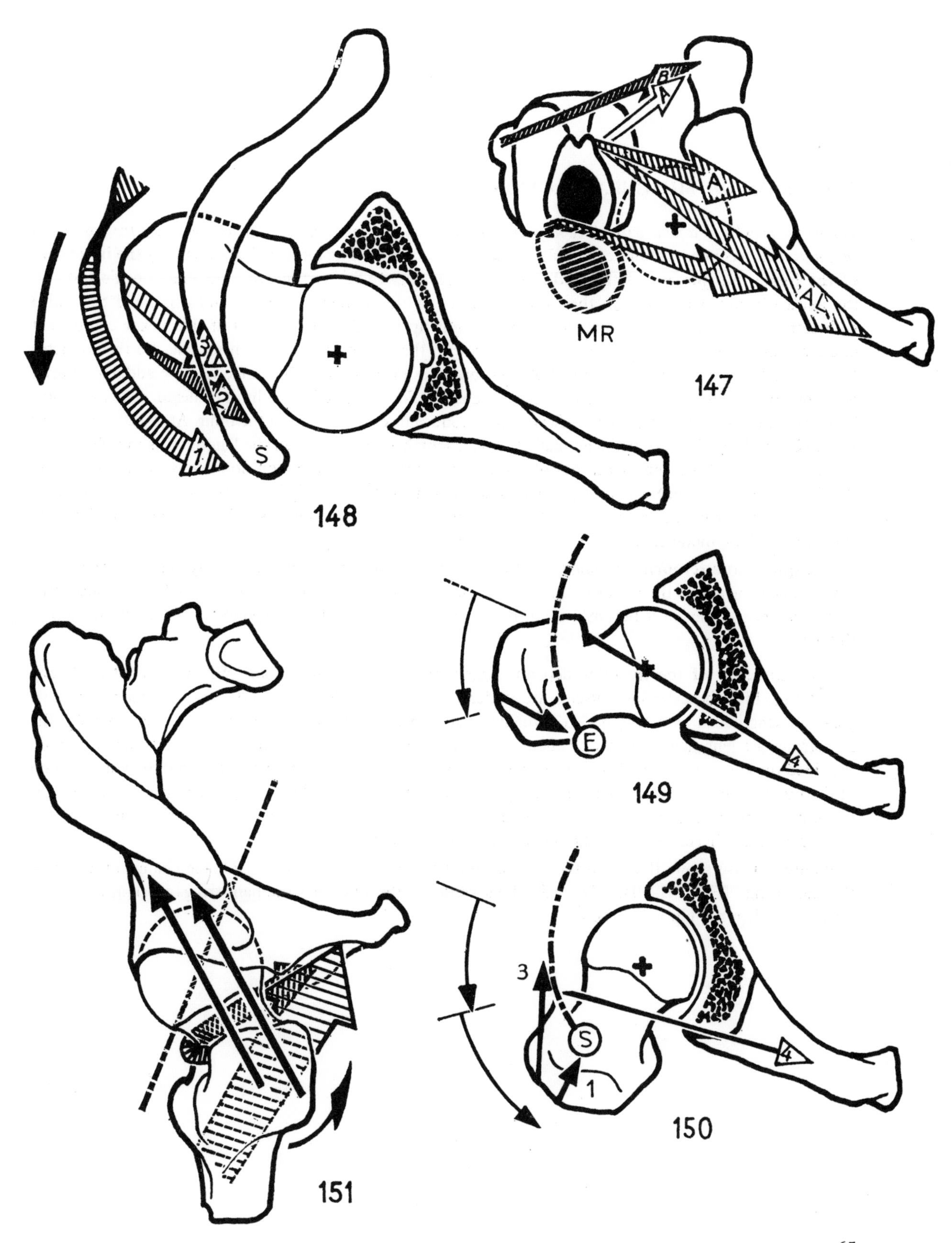

B
A
A
AL
MR
147
3
2
1
S
148
E
4
149
3
S
1
4
150
151

INVERSION OF MUSCULAR ACTION

The motor muscles of a joint with three degrees of freedom do not have the same action whatever the position of the joint; their secondary actions can be altered and even reversed. The most typical example is **the inversion of the flexor component of the adductor muscles** (fig. 152). Starting from the erect position (0°) all the adductors are flexors except the posterior fibres of the adductor magnus and especially the 'third adductor' (TA), which is and remains extensor right up to −20° extension. But this flexor component operates only so long as the femur still lies inferior to the site of origin of each muscle. So the adductor longus (AL) is still flexor at position +50° but becomes extensor at +70°. Likewise the adductor brevis is flexor up to +50°, after which it produces extension. For the gracilis the limit of flexor action is +40°. The diagram shows that only the true flexors can produce flexion right up to the limit. At +120° the tensor fasciae latae (TFL) is shortened maximally (i.e. by a length aa′ equal to half the length of its muscle fibres), while the psoas (P) has almost run out of useful contraction since its tendon now tends to 'take off' from the iliopectineal eminence. (The diagram suggests an explanation of 'why' the lesser trochanter is situated very far posteriorly: the excursion of the psoas tendon is thereby increased by a length equal to the thickness of the femoral shaft.)

The **quadratus femoris** also shows this inversion of muscular action very clearly (fig. 153: the transparent iliac bone allows the femur and the quadratus to be seen): in extension (E) of the hip the quadratus is flexor, while in flexion it becomes extensor; the point of transition corresponds to the erect position of the hip.

The **efficiency of the muscles** depends largely on *the position of the hip*. When the joint is already flexed (fig. 154) the extensor muscles are under tension. With flexion of 120° the gluteus maximus is passively lengthened by a length GG′, which for certain fibres represents 100 per cent. lengthening; the hamstrings are lengthened (HH′) by about 50 per cent. of their length in the 'straight' position of the hip provided the knee stays extended. This explains the starting position of runners (fig. 155): maximal flexion of the hip followed by extension of the knee (this second phase is not illustrated), which produces the right amount of tension in the hip extensors for the powerful movement of the start. It is this tension of the hamstrings that checks flexion of the hip when the knee is extended.

Figure 154 also shows that from the erect position to that of extension at −20° the change in length of the hamstrings (HH″) is relatively small. This confirms the idea that the hamstrings work at their best advantage when the hip is half-flexed.

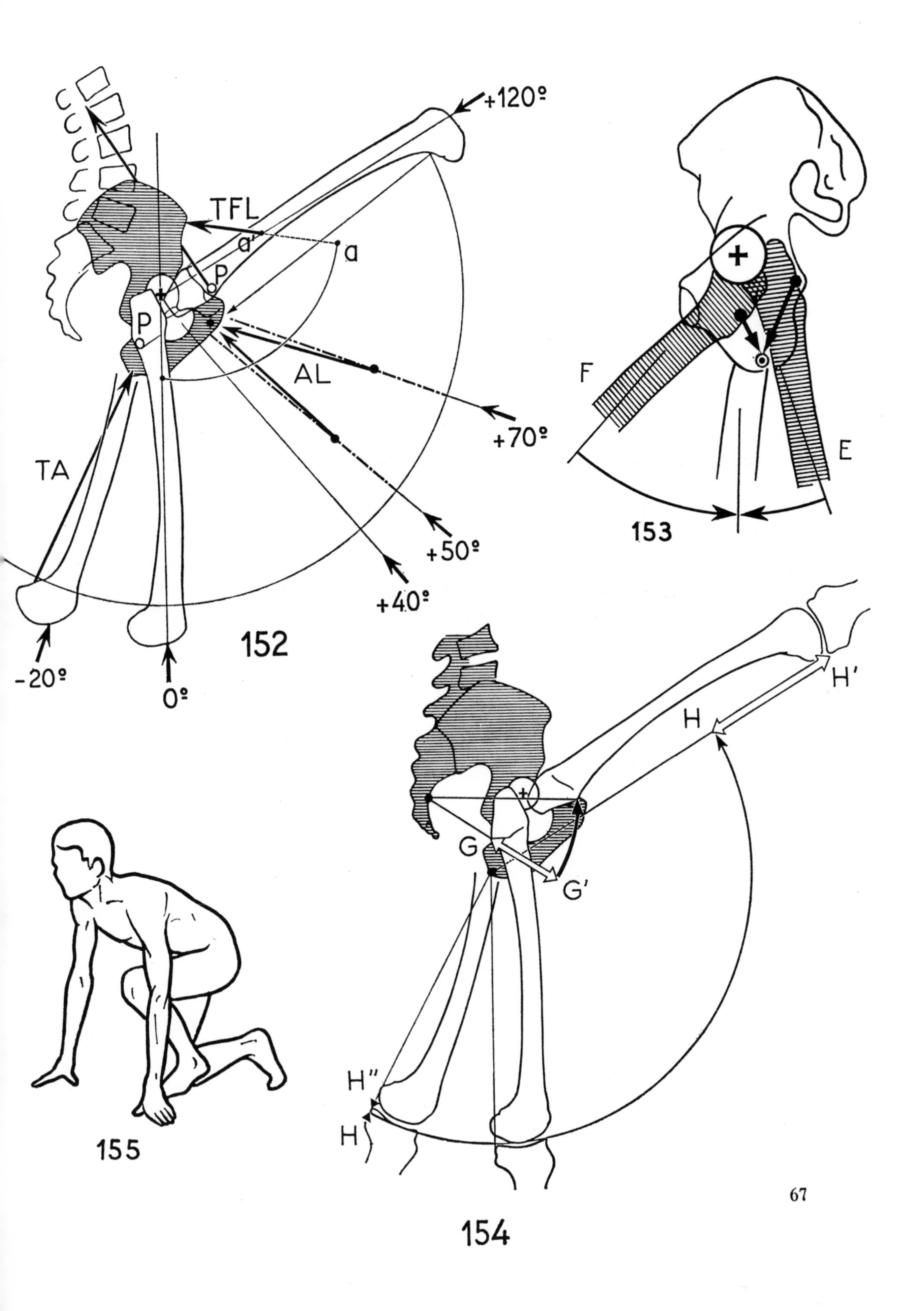
+120º
TFL
a'
a
P
P
AL
+70º
TA
+50º
+40º
-20º
0º
152
F
E
153
H'
H
G
G'
H"
H'
155
154

When the hip is in a position of exaggerated flexion (fig. 156) the piriformis also shows inversion of its actions. When the hip is in the 'straight' position (fig. 157) it produces lateral rotation, *flexion* and abduction white arrow); when flexion is very marked it produces medial rotation, *extension*, abduction (striped arrow). The point of inversion corresponds to 60° flexion, at which point it is only an abductor. In marked flexion again (fig. 158: the flexed hip seen from behind and from outside) abduction is produced not only by the piriformis (P) but also by the obturator internus (OI) and all the fibres of gluteus maximus (GMa). These muscles produce separation of the two knees when the hip is flexed at 90°. The gluteus minimus (GMi) is quite clearly a medial rotator and becomes an adductor (fig. 159) along with the tensor fasciae latae (T) and the resultant movement produced by these muscles has the following three components: flexion, adduction, medial rotation (fig. 160).

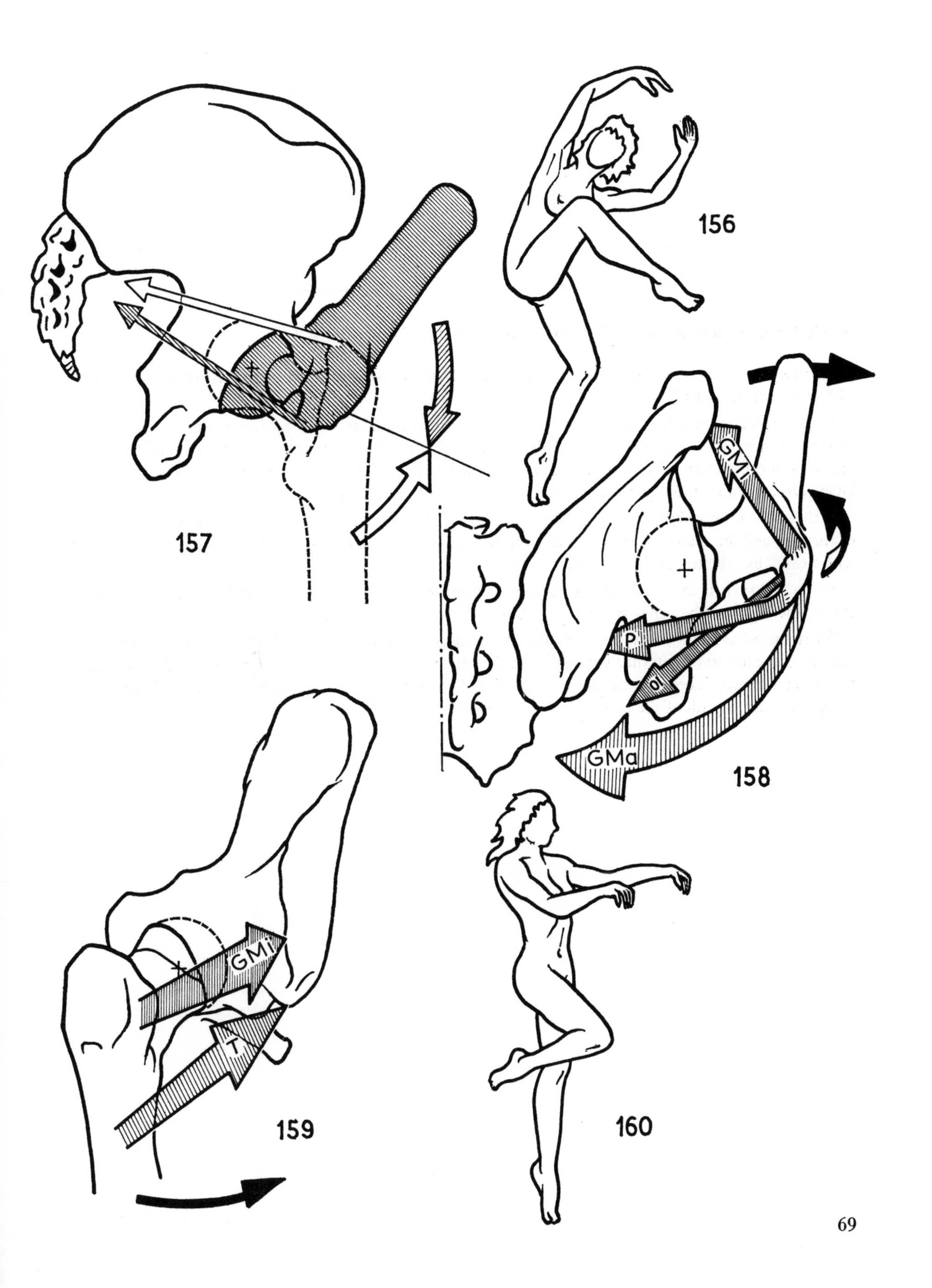
156
157
GMi
P
a
GMa
158
GMi
T
159
160

SUCCESSIVE RECRUITMENT OF THE ABDUCTOR MUSCLES

Depending on the degree of flexion of the hip, the pelvis, **when supported on one limb**, is stabilised by different abductor muscles.

When the hip is extended (fig. 161), the centre of gravity of the body falls posterior to the transverse axis of the two hips. The posterior tilting of the pelvis is checked by the tension of the iliofemoral ligament (p. 36) and contraction of the tensor fasciae latae which is also a flexor of the hip. Hence the tensor fasciae latae corrects simultaneously posterior and lateral tilting of the pelvis.

If the pelvis is only slightly tilted posteriorly (fig. 162) the centre of gravity still lies posterior to the axis of the hips and the gluteus minimus is thrown into action. Note that this muscle also produces abduction with flexion like the tensor fasciae latae.

When the pelvis is in equilibrium in the anteroposterior plane (fig. 163) the centre of gravity lies on the axis of the hips and the pelvis is laterally stabilised by the gluteus medius.

When the pelvis is tilted forward the gluteus maximus is called into action and subsequently the piriformis (fig. 164), the obturator internus (fig. 165) and the quadratus femoris (fig. 166), are recruited as flexion of the trunk increases. These muscles are also *abductors* (when the hip is flexed) and *extensors*, so that they can compensate for any tilt of the pelvis in the anteroposterior and transverse planes.

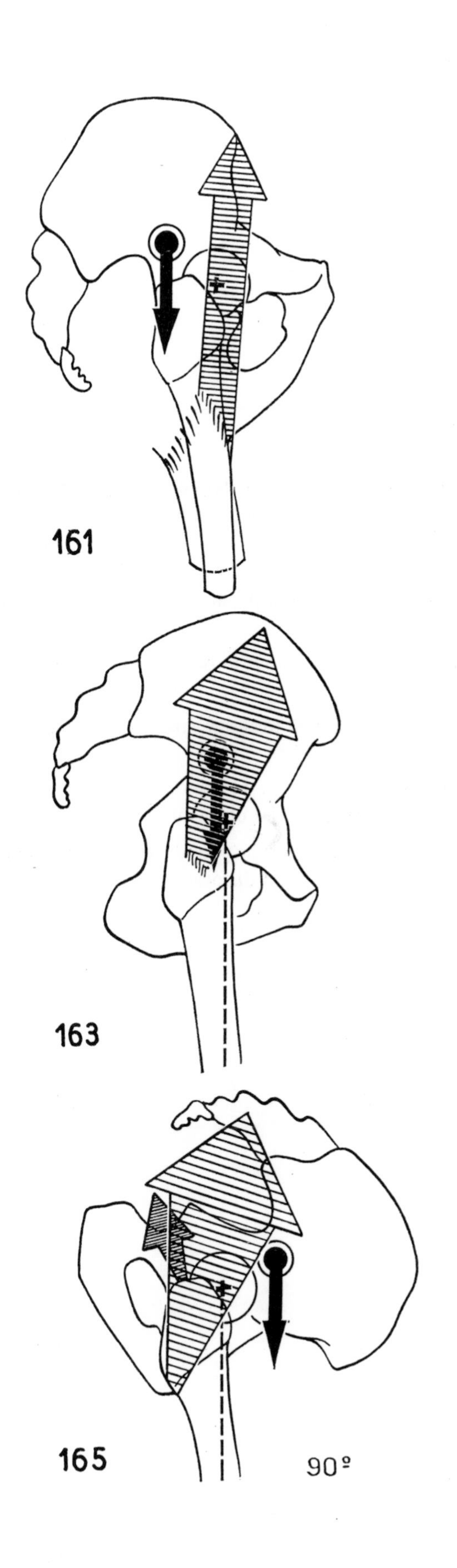
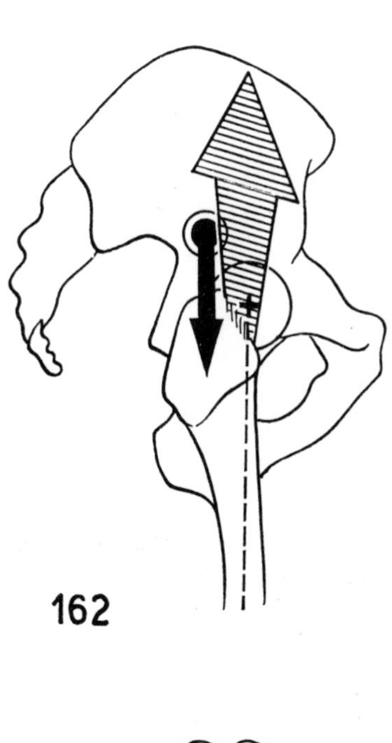

161

162

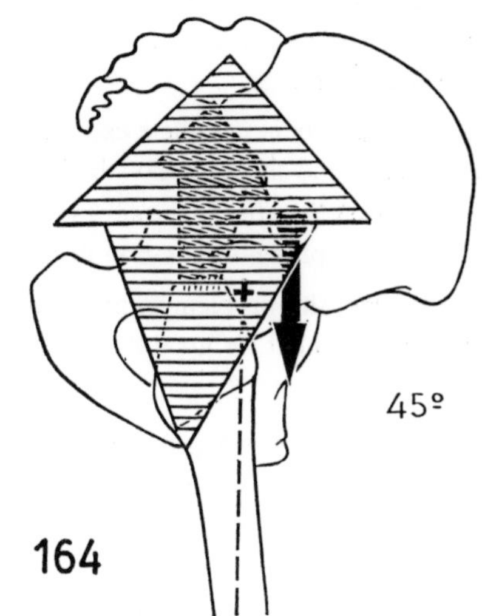

163

164

45º

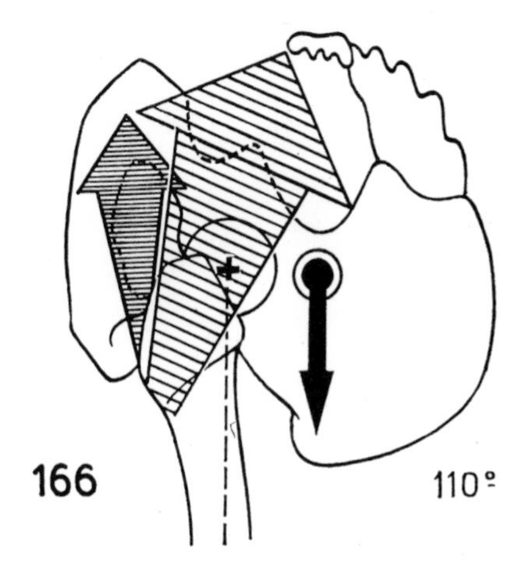

165

90º

166

110º

THE KNEE

The knee is the *intermediate* joint of the lower limb. It is *mainly* a joint with **one degree of freedom** which allows the end of the limb to be moved towards or away from its root or, in other words, allows the distance between the trunk and the ground to be varied. *The knee works essentially by axial compression* under the action of gravity.

The knee has an **accessory, i.e. second, degree of freedom**: rotation of the long axis of the leg, which only occurs *when the knee is flexed*.

From the mechanical point of view the knee is a compromise which sets out to reconcile **two mutually exclusive requirements**:

to have *great stability* in complete extension, when the knee is subjected to severe stresses resulting from the body weight and the length of the lever arms involved;

to have *great mobility* after a certain measure of flexion has been achieved. This mobility is essential for running and the optimal orientation of the foot relative to the irregularities of the ground.

The knee resolves this problem by highly ingenious mechanical devices but the poor degree of interlocking of the surfaces—essential for great mobility—renders it liable to sprains and dislocations.

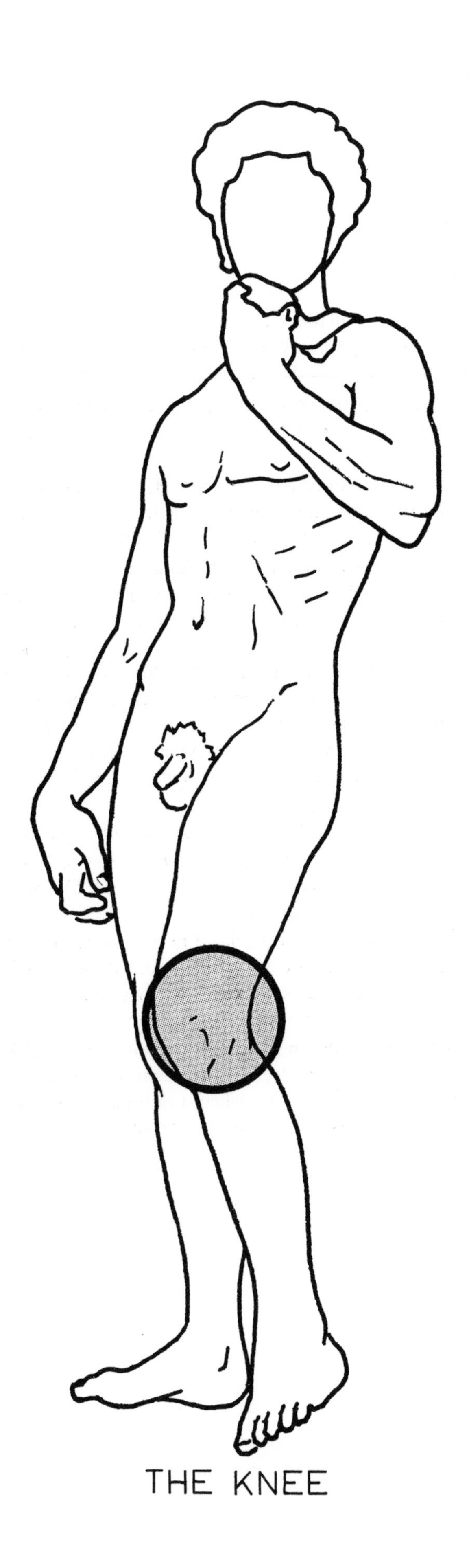

THE KNEE

THE AXES OF THE KNEE

The **first degree of freedom** is related to the *transverse axis* XX′ (fig. 1, semiflexed knee seen from inside; fig. 2, seen from outside), around which occur movements of flexion and extension in a sagittal plane. This axis XX′, lying in a frontal plane, runs through the femoral condyles horizontally.

Because the femoral neck overhangs the shaft (fig. 3), the axis of the femoral shaft does not coincide with that of the leg but forms with the latter an obtuse angle of 170 to 175° opening outwards. This is the *physiological valgus of the knee.*

On the other hand, the centres of the three joints i.e. hip (H), knee (O), ankle (C), lie on a straight line HOC which is the **mechanical axis** of the lower limb. In the leg it coincides with that of the leg itself but in the thigh it forms an acute angle of 6° with the axis of the femur.

Because the hips are wider apart than the ankles, *the mechanical axis of the lower limb runs obliquely inferiorly and medially* and forms an angle of 3° with the vertical. This angle is greater the wider the pelvis, as in the case of women. This also explains why the physiological valgus of the knee is *more marked in women* than in men.

The axis of flexion and extension XX′ is horizontal and so does not coincide with Ob which bisects the angle of valgus. The angle between XX′ and the femoral axis is 81° and that between XX′ and the axis of the leg is 93°. Therefore during full flexion the axis of the leg does not come to rest immediately posterior to that of the femur but *posterior and slightly medial to it* so that the heel moves medially towards the plane of symmetry of the body. Full flexion brings *the heel into contact with the buttock at the level of the ischial tuberosity.*

In addition to these sex-linked physiological variations, the angle of valgus shows pathological variations. When the angle is reversed the condition is called *genu varum* (fig. 4) i.e. 'bandy legs'. When the angle of valgus is exaggerated this is called *genu valgum,* i.e. 'knock-kneed' (fig. 5). Genu valgum is fairly common in the toddler but generally disappears with growth.

The second degree of freedom of the joint is related to rotation around the *long axis* YY′ of the leg (figs. 1 and 2) *with the knee flexed.* The structure of the knee makes **axial rotation impossible when the knee is fully extended**; the axis of the leg coincides with the mechanical axis of the lower limb and axial rotation occurs not at the knee but at the hip, which is thus complementary to the knee.

In figure 1 an axis ZZ′ (broken line) is shown running anteroposteriorly and at right angles to the other two axes. This axis does not represent a third degree of freedom but, owing to a measure of 'play' at the joint, **side-to-side movements** occur (1 to 2 cm. when measured at the ankle). In full extension these movements disappear and if they still persist they must be considered, as a rule, as *abnormal.*

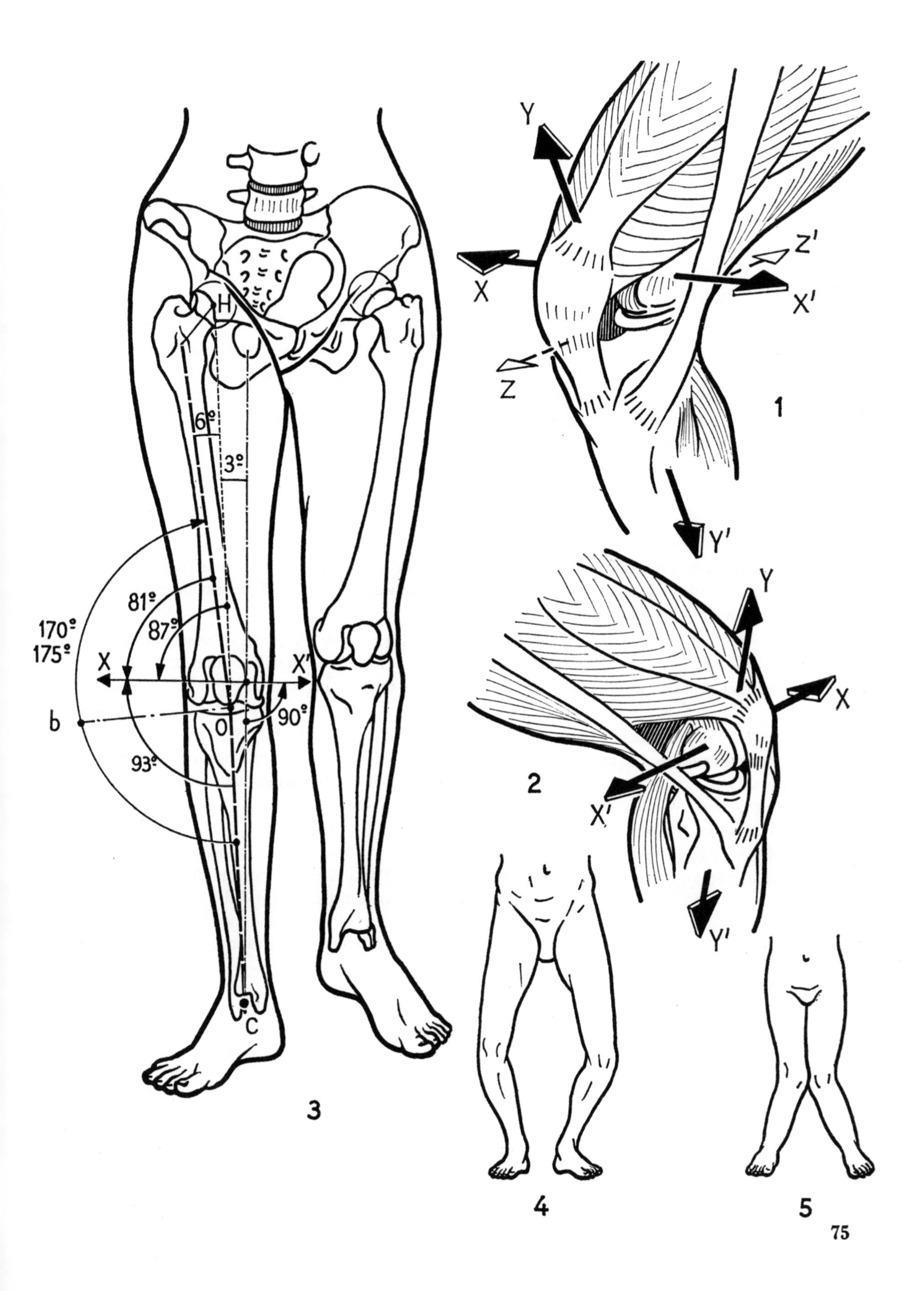

Y
X
Z
Z'
X'
1
Y'
Y
X
2
X'
Y'
6°
3°
81°
87°
170°
175°
X
X'
b
O
90°
93°
H
C
3
4
5

MOVEMENTS OF THE KNEE AND THEIR RANGES: FLEXION AND EXTENSION

These are the main movements of the knee and their range is measured from the **position of reference** established by the following criterion: *the axis of the leg is in line with that of the thigh* (fig. 6, left leg), i.e. seen from the side, the axis of the femur prolongs that of the leg directly. In this position of reference, the lower limb is at its longest.

Extension is defined as the movement of the posterior aspect of the leg *away* from the posterior surface of the thigh. There is strictly no **absolute extension** since in the position of reference the limb is maximally extended. It is, however, possible to achieve *passive extension* (5° to 10°) from the position of reference (fig. 8): this is wrongly called 'hyperextension'. In some people this hyperextension can be abnormally marked leading to the *genu recurvatum.*

Active extension goes beyond the position of reference rarely and then only slightly (fig. 6) and this depends upon the position of the hip joint. In fact, the efficiency of the rectus femoris as an extensor of the knee increases with extension of the hip (p. 128), so that extension of the hip (fig. 7: right leg) sets the stage for knee extension.

Relative extension is the movement which brings the knee into full extension starting from any position of flexion (fig. 7: left leg). It occurs normally during walking when the limb off the ground is extended to resume contact with the ground (Vol. IV).

Flexion is the movement of the posterior aspect of the leg *towards* the posterior aspect of the thigh. Flexion can be *absolute*, i.e. from the reference position, and *relative*, i.e. from any position of partial flexion.

The range of knee flexion varies according to the position of the hip and according to whether it is active or passive.

Active flexion attains a range of 140° if the hip is already flexed (fig. 9) and only 120° if the hip is extended (fig. 10). This difference is due to the fact that the hamstrings lose some of their efficiency with extension of the hip (p. 130). It is, however, possible to exceed 120° flexion with the hip extended as a result of a '*follow-through*' *effect*. When the hamstrings contract abruptly and powerfully the knee is thrown into flexion and the end of active flexion is followed by a measure of passive flexion.

Passive flexion of the knee attains a range of 160° (fig. 11) and *allows the heel to touch the buttock*. This movement underlies an important clinical test of the freedom of flexion of the knee and the range of passive flexion of the knee can be assessed in terms of the distance between the heel and the buttock. Normally, flexion is checked only by the apposition of the elastic muscle masses of the calf and the thigh. Pathologically, passive flexion is limited by *retraction of the extensor apparatus*—essentially the quadriceps—or by *shortening of the capsular ligaments* (p. 102).

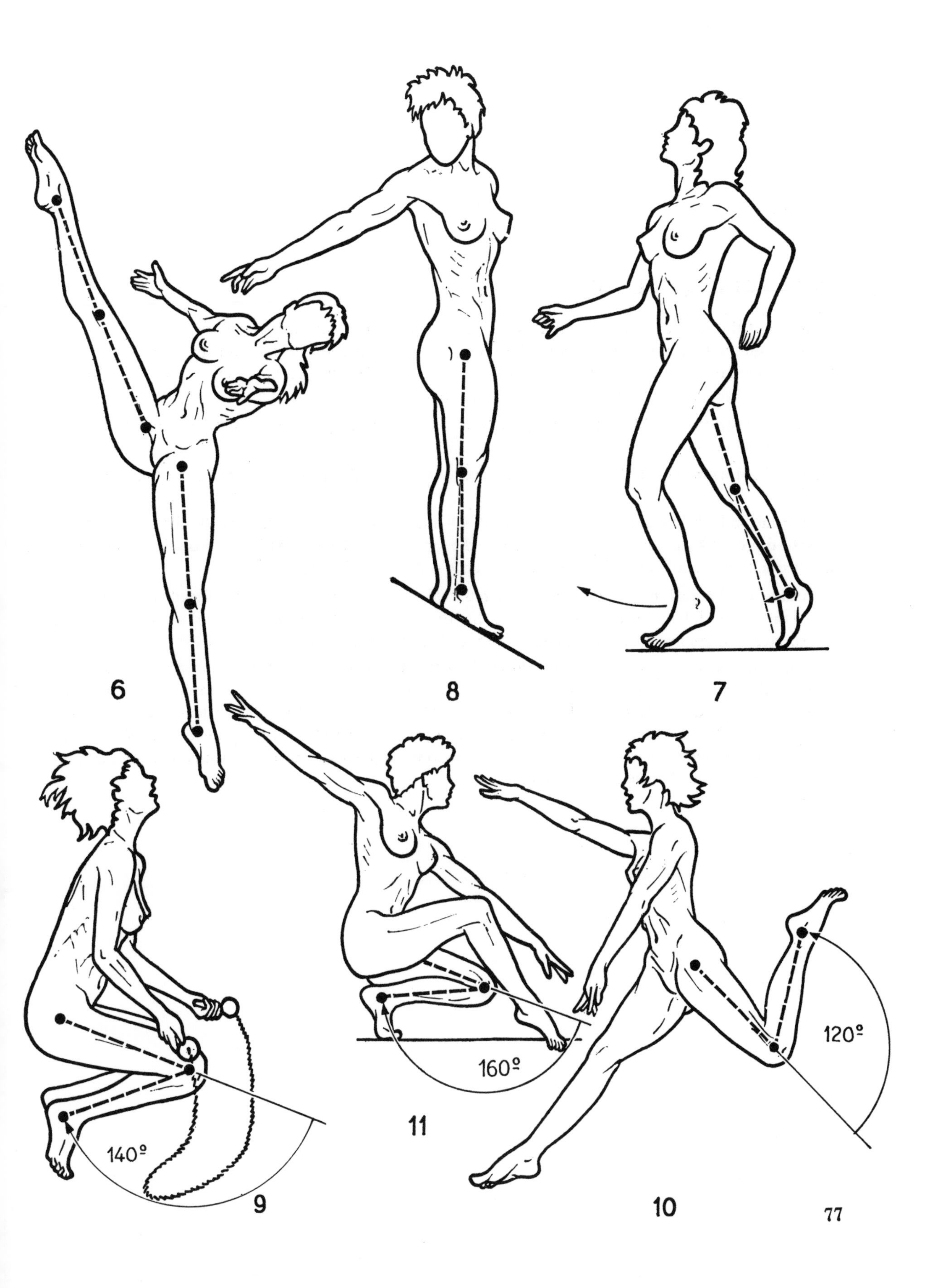
6
8
7
9
140º
11
160º
10
120º

AXIAL ROTATION OF THE KNEE

Rotation of the leg around its long axis can only be performed with the knee flexed.

To measure **active axial rotation** the knee must be flexed at right angles with the subject sitting at the edge of a table with his legs hanging down (fig. 12); knee flexion prevents rotation at the hip. In the position of reference the toes face slightly outwards (p. 80).

Medial rotation (fig. 13) brings the toes to face *medially* and plays an important part in adduction of the foot (p. 156).

Lateral rotation (fig. 16) brings the toes to face *laterally* and also plays an important part in abduction of the foot.

According to Fick, lateral rotation has a range of 40° and medial rotation a range of 30°. This range varies with the degree of knee flexion since, according to the same author, lateral rotation attains a 32° range when the knee is flexed at 30° and 42° range when the knee is flexed at right angles.

Passive axial rotation can be measured when the subject lies prone with the knee flexed. The examiner grasps the foot with both hands and turns it so that the toes face outwards (fig. 15) and inwards (fig. 16). As expected, this passive rotation has a greater range than active rotation.

Finally there is also a type of **axial rotation called automatic** because it is inevitably and involuntarily linked to movements of flexion and extension. It occurs especially at the end of extension or the start of flexion. When the knee is extended the foot is *laterally* (*EXTernally*) *rotated* (fig. 17); hence the mnemonic EXTension and EXTernal rotation. Conversely, when the knee is flexed the leg is *medially rotated* (fig. 18). The same movement takes place when one tucks the lower limbs underneath the trunk and the toes are brought to face medially (the foetal position).

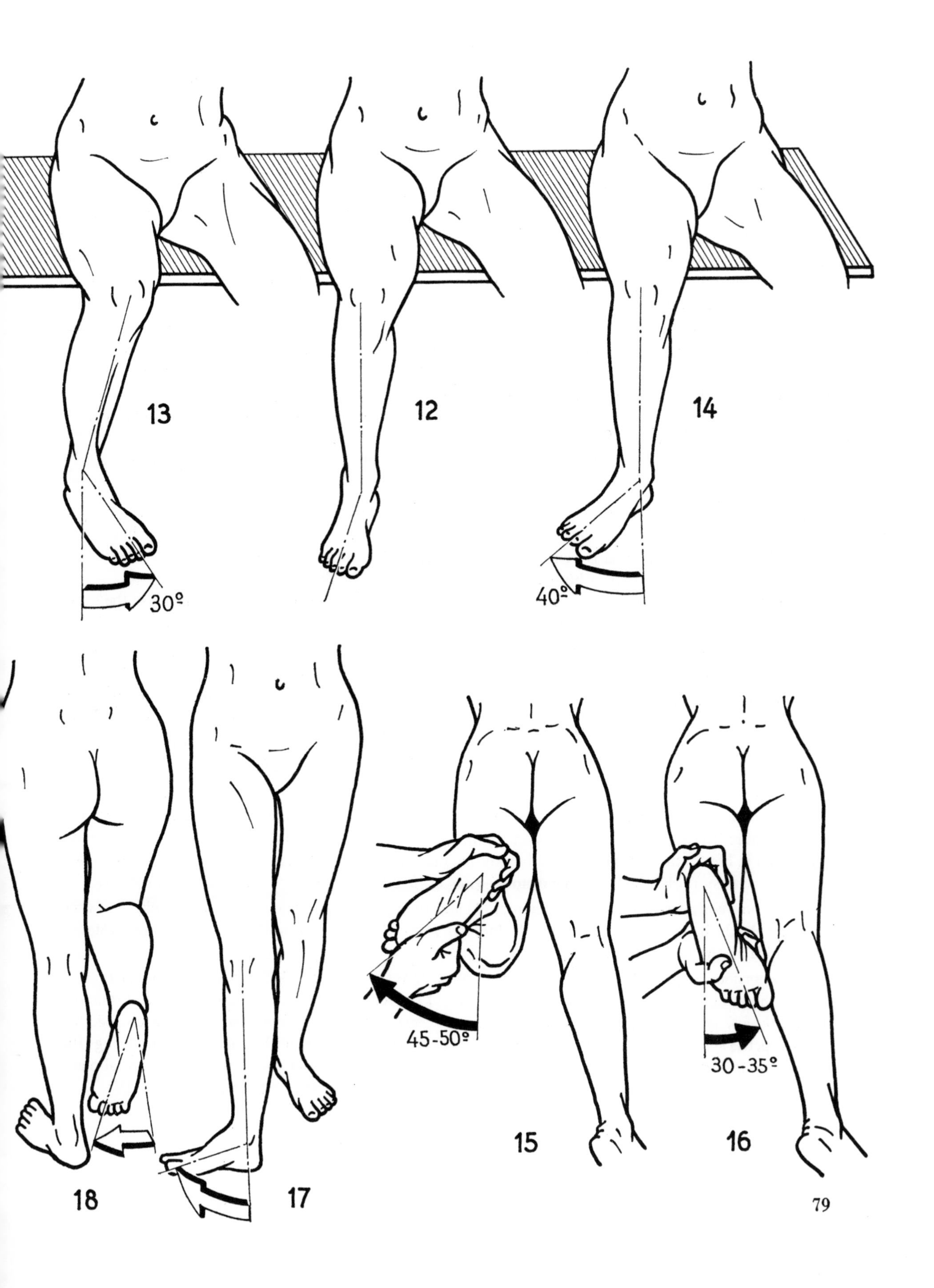
13
12
14
30º
40º
45-50º
30-35º
15
16
18
17

THE GENERAL STRUCTURE OF THE LOWER LIMB AND THE ORIENTATION OF THE ARTICULAR SURFACES

The orientation of the femoral condyles and the tibial condyles favours flexion of the knee (fig. 19, according to Bellugue). Two bony extremities, moving relative to each other (a), become moulded according to their movements (b) (experiment of Fick, Vol. III). Flexion falls short of a right angle (c) unless a small fragment (d) is removed from the upper bone so as to delay contact of the two bones. The weak point thus created in the bone is compensated by forward displacement (e) of the shaft so that the condyles come to lie posteriorly. Reciprocally, the tibia is thinned posteriorly and reinforced anteriorly (f) so that the tibial plateau comes to lie more posteriorly. Thus, during extreme flexion, large muscular masses can be lodged between tibia and femur.

The overall curvatures of the lower limb reflect the stresses which are applied and obey the laws of Euler governing the behaviour of *columns eccentrically loaded* (Steindler). If a column is jointed at its two ends (fig. 20, a) the bend involves the whole length of the column; hence the bend of the femur, concave posteriorly (fig. 20, b). If the column is fixed below and mobile above (fig. 21, a) two bends with opposite curvatures are seen, with the higher bend taking up two-thirds of the column; these bends correspond to those of the femur in the frontal plane. If the column is fixed at both ends (fig. 22, a) the bend takes up its two middle fourths; this corresponds to the bends of the tibia in the frontal plane (fig. 22, b). In the sagittal plane the tibia shows the following three features (fig. 23, b):

retrotorsion (T), i.e. posterior displacement of the upper end, already mentioned;
retroversion (V), i.e. the tibial condyles are inclined posteriorly at an angle of 5° to 6° with the horizontal;
retroflexion (F), i.e. the tibia is bent so as to be concave posteriorly; this corresponds to the bend seen in a column which is mobile at both ends (fig. 23, a), as with the femur.

During flexion (fig. 24) the concavities of the femur and tibia face each other and so increase the space for lodgement of the soft tissue masses.

The diagrams at the bottom of the page constitute a sort of 'anatomical algebra', attempting to explain *the successive movements of axial torsion of the bones of the lower limb*, as seen from above.

Torsion of the femur (fig. 25): let us assume that (a) the head and neck (1) and the condyles (2) constitute a single solid structure. If no torsion is present (b) the axis of the neck lies in the same plane as that of the condyles. However in life the neck forms an angle of 30° with the frontal plane (c): therefore, if the axis of the condyles is to stay in the frontal plane (d), the *femoral shaft must be twisted by −30° of medial rotation.*

Torsion of the tibia (fig. 26): let us assume (a) that the ankle (1) and the tibial condyles (2) form one solid structure. In the absence of any torsion (b) the axis of the condyles and that of the ankle lie in a frontal plane. But in life (c) the retroposition of the external malleolus causes the axis of the ankle to run obliquely laterally and posteriorly; this corresponds to a *torsion of the tibia equivalent to +25° of lateral rotation.*

Let us now consider (fig. 27, a) the femoral (1) and the tibial (2) condyles together. It looks as if their two axes should lie in a frontal plane (b). In life, *automatic axial rotation* produces +5° of lateral rotation of the tibia on the femur during full extension of the knee.

These torsions, staggered over the whole length of the lower limb (−30°+25°+5°), cancel out (fig. 28, a) so that the axis of the ankle is roughly parallel to that of the femoral neck, i.e. showing 30° lateral rotation. Hence *the axis of the foot is set at 30° to the plane of symmetry* when one stands erect with heels together and the pelvis symmetrically balanced (b). During walking the movement of the advancing limb brings the ipsilateral hip forwards (c); if the pelvis turns by 30°, the axis of the foot comes to lie in a sagittal plane i.e. in the plane of movement and this allows the step to evolve under the best conditions (Vol. IV).

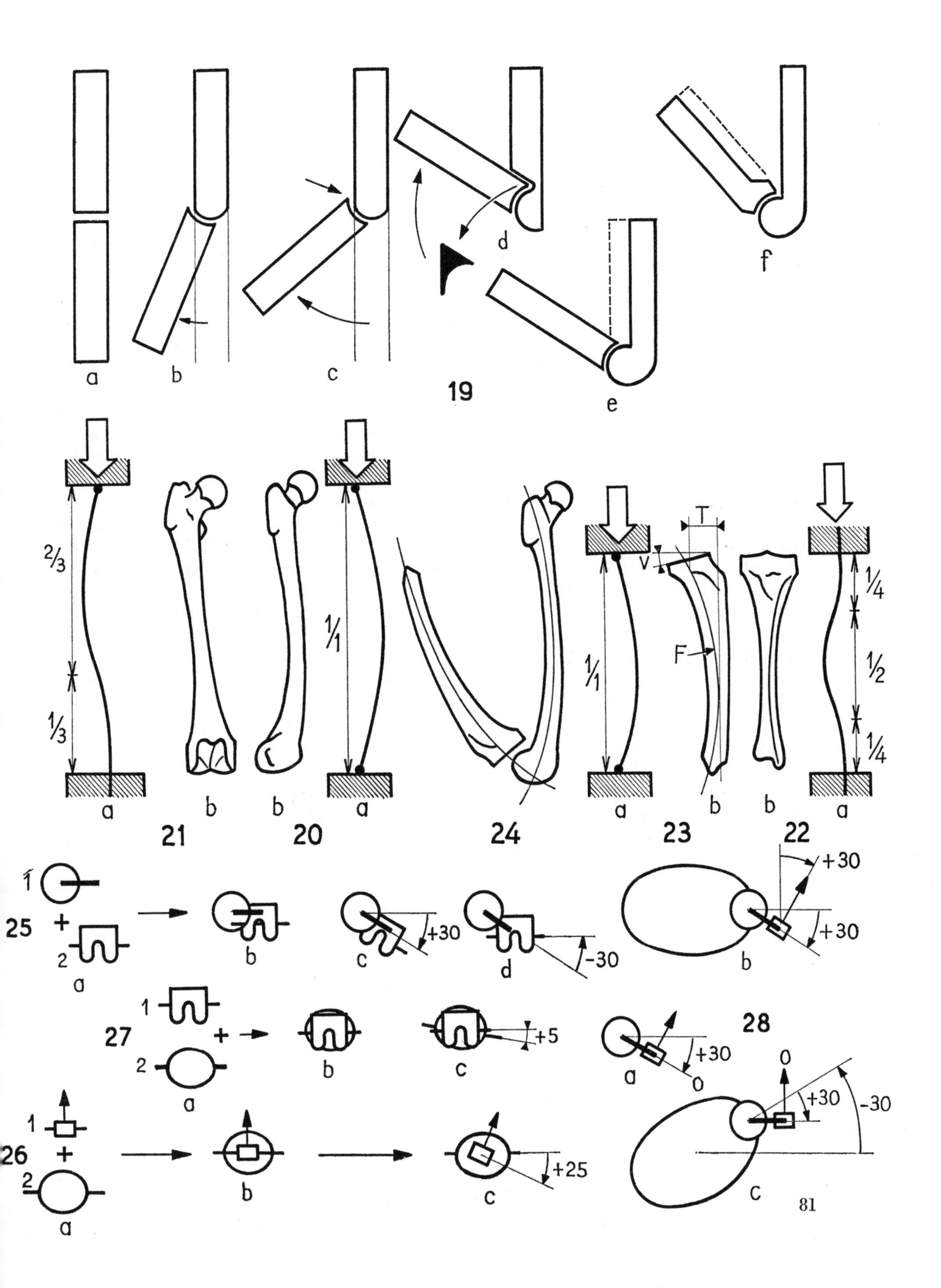

a
b
c
19
d
e
f
2/3
1/3
a
b
21
b
20
1/1
a
24
1/1
a
T
V
F
b
23
b
22
1/4
1/2
1/4
a
1
25
+
2
a
b
c
+30
d
-30
+30
+30
b
1
27
+
2
a
b
c
+5
28
a
+30
0
0
1
26
+
2
a
b
c
+25
+30
-30
c

ARTICULAR SURFACES INVOLVED IN FLEXION AND EXTENSION

The main movement of the knee, i.e. flexion and extension, occurring around a transverse axis, depends upon the fact that it is a **hinge joint**. In fact, the articular surfaces of the femur are *pulley-shaped*, or, more exactly, represent a segment of a pulley (fig. 29), which recalls, in a general way, the *twin undercarriage of an aeroplane* (fig. 30). The two femoral condyles, convex in both planes, form the two lips of the pulley and correspond to the wheels of the aeroplane; they are extended anteriorly (fig. 31) by the pulley-shaped patellar surface. The neck of the pulley is represented anteriorly by the central groove on the patellar surface and posteriorly by the intercondylar notch (the mechanical significance of this arrangement will emerge later). Some authors describe the knee as a bicondyloid joint: this is true anatomically but from the mechanical point of view the joint is indisputably of the hinge variety.

The tibial surfaces are reciprocally curved and comprise *two curved and concave parallel gutters which are separated by a blunt eminence running anteroposteriorly* (fig. 32). The lateral condyle (LC) and the medial condyle (MC) lie each in a gutter on the surface S and are separated by a blunt anteroposterior eminence which lodges the two intercondylar tubercles. Anteriorly, if this eminence is prolonged, it coincides with the vertical ridge on the deep surface of the patella (P), while the two facets on either side of the patellar ridge correspond to the tibial condyles. These surfaces have a transverse axis (I) which coincides with the intercondylar axis (II) when the joint is closed.

Therefore the tibial condyles correspond to the femoral condyles while the intercondylar tibial tubercles come to lie within the femoral intercondylar notch: these surfaces constitute functionally **the femorotibial joint**. Anteriorly the two facets of the patella correspond to the patellar surface of the femur while the vertical ridge of the patella fits into the central groove of the femur: these surfaces constitute the second *functional* joint, i.e. the **femoropatellar joint**. These two functional joints are contained within a single *anatomical* joint, i.e. the knee.

From the point of view of flexion and extension, one can in the first instance consider the knee as made up of a pulley-shaped surface gliding in a twin set of curved and concave gutters (fig. 33).

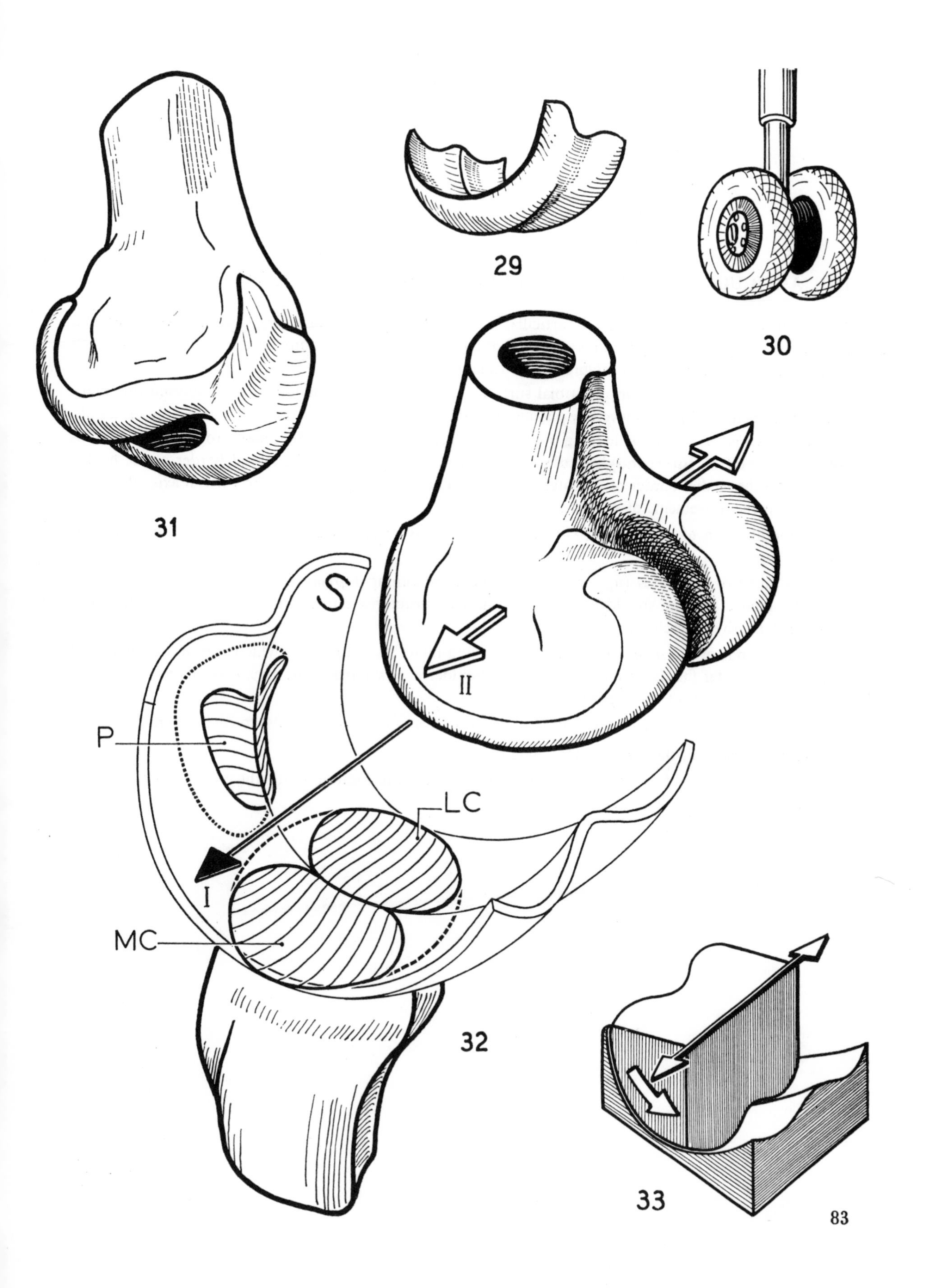
29
30
31
S
II
P
LC
I
MC
32
33

THE ARTICULAR SURFACES IN RELATION TO AXIAL ROTATION

As described on the preceding page, the articular surfaces allow only one movement, i.e. flexion and extension. In fact, the tibial intercondylar eminence by lodging in the intercondylar notch of the femur *along its whole length* would preclude any axial rotation of the tibia on the femur.

Hence, to allow axial rotation, the tibial surface (fig. 34) must be so modified as to shorten the intercondylar eminence. This is achieved by planing (fig. 35) the two ends of the eminence and leaving its middle part to act as a *pivot*, which, by lodging in the intercondylar notch, allows the tibia to rotate round it. This pivot consists of the **intercondylar tubercles** which form respectively the lateral border of the medial condyle and the medial border of the lateral condyle. Through this point runs the vertical axis (R), about which occur the movements of axial rotation.

This modification of the articular surfaces is more easily understood with the help of a **mechanical model.**

Let us take two structures (fig. 36): the one above containing a *groove* and the one below a *shoulder*. Since the groove and the shoulder fit exactly, the two structures can *slide* one over the other but they *cannot turn* relative to each other.

If the two ends of the shoulder are removed leaving intact its central part which is rounded off so that its greatest diameter fits the groove (fig. 37), the shoulder is now replaced by a *cylindrical knob*, which can still fit into the groove of the upper surface.

Now (fig. 38) these two structures can undergo *two types of movement* relative to each other:

a sliding movement of the central knob along the groove

a movement of rotation of the knob inside the groove (whatever its location within the groove), corresponding to rotation around the long axis of the leg.

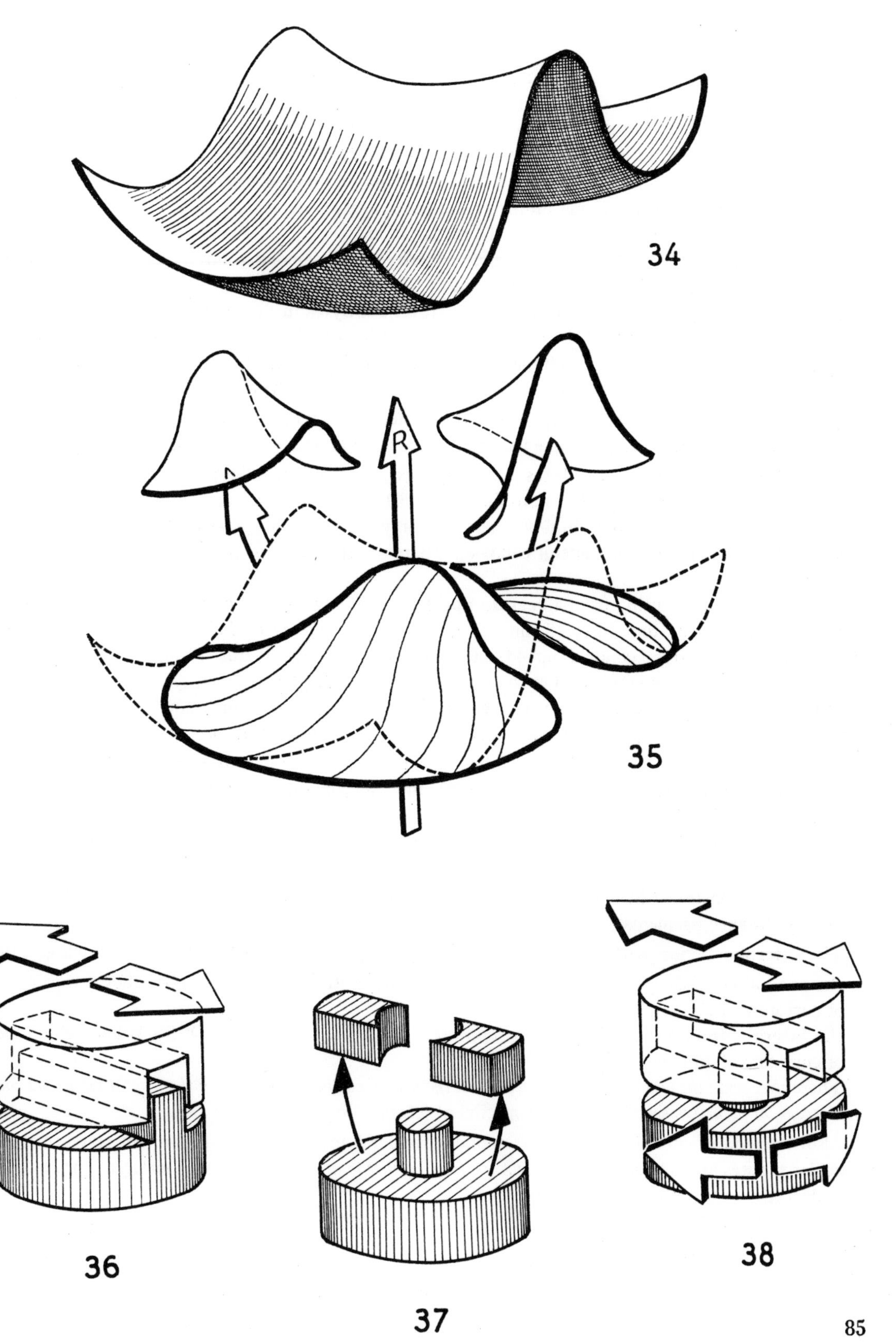
34
R
35
36
37
38

PROFILE OF THE FEMORAL AND TIBIAL CONDYLES

When seen from below (fig. 39) the **femoral condyles** form two prominences *convex in both planes* and longer anteroposteriorly than transversely. They are not strictly identical: their long axes (i.e. antero-posterior) are not parallel but *diverge posteriorly*. Moreover, the medial condyle (M) juts out far more than the lateral condyle (L) and is also narrower. Between the patellar surface of the femur and the condyles run the medial and lateral grooves (r), the medial one being more obvious than the lateral.

The **intercondylar notch** (e) lies on the axis of the *central groove of the patellar surface* (g). The lateral facet of this surface is *more prominent* than the medial facet.

The frontal section (fig. 40) shows that the convexity of the femoral condyles in the transverse plane corresponds to the concavity of the tibial condyles.

To study the **curvature of the femoral and tibial condyles** in the sagittal plane it is convenient to examine sagittal sections taken at the levels aa′ and bb′ (fig. 40). These sections taken from a fresh bone provided a faithful profile of the femoral and tibial condyles (figs. 42 to 45). It is then clear that the radius of curvature of the femoral condyles is not uniform but varies as in the case of a **spiral.**

In geometry, Archimedes' spiral (fig. 41) is constructed around a point called its centre C so that every time the radius R sweeps over an angle its length is correspondingly increased.

The **spiral of the femoral condyles** is *quite different*, though their radii of curvature increase regularly postero-anteriorly, i.e. from 17 to 38 mm. for the medial condyle (fig. 42) and from 12 to 60 mm. for the lateral condyle (fig. 43). But the spiral does not have only one centre but *a series of centres lying on a spiral mm′* (medial condyle) and nn′ (lateral condyle). Therefore the curve of the condyles represents a *spiral of a spiral.*

On the other hand, starting from a **certain point t** on the edge of the condyle, the radius of curvature begins to decrease, i.e. from 38 to 15 mm. postero-anteriorly for the medial condyle (fig. 42) and from 60 to 16 mm. for the lateral condyle (fig. 43). Here too the centres of curvature lie on a spiral m′m″ (medial condyle) and n′n″ (lateral condyle). On the whole, *the lines joining these centres of curvature constitute two spirals lying back to back* and having a very sharp apex (m′ and n′), which corresponds to the point t on the condyle, i.e. the point of transition between *the two segments of the condylar profile*:

posterior to point t, the part of the condyle belonging to the *femorotibial joint*;

anterior to point t, the part of the condyle and patellar surface of the femur belonging to the *femoropatellar joint.*

The transition point t represents the most extreme point of the condyle able to come into contact with the tibia.

We have demonstrated by a mechanical model that the trochleo-condylar profile is geometrically determined by the knee ligaments.

The **anteroposterior profile of the tibial condyles** (figs. 44 and 45) varies with the condyle examined.

the medial condyle (fig. 44) is *concave superiorly* (the centre of curvature O lies above) with a radius of curvature of 80 mm;

the lateral condyle (fig. 45) is *convex superiorly* (the centre of curvature O′ lies below) with a radius of curvature of 70 mm.

Therefore, *while the medial condyle is biconcave, the lateral condyle is concave in the frontal plane and convex in the sagittal plane* (as seen in the fresh specimen) (cf. fig. 35).

Also, the radii of curvature of the corresponding femoral and tibial condyles are not equal so that the articular surfaces are not congruent. In fact *the knee typifies joints with incongruent surfaces.* The restoration of congruence rests with the menisci (p. 96).

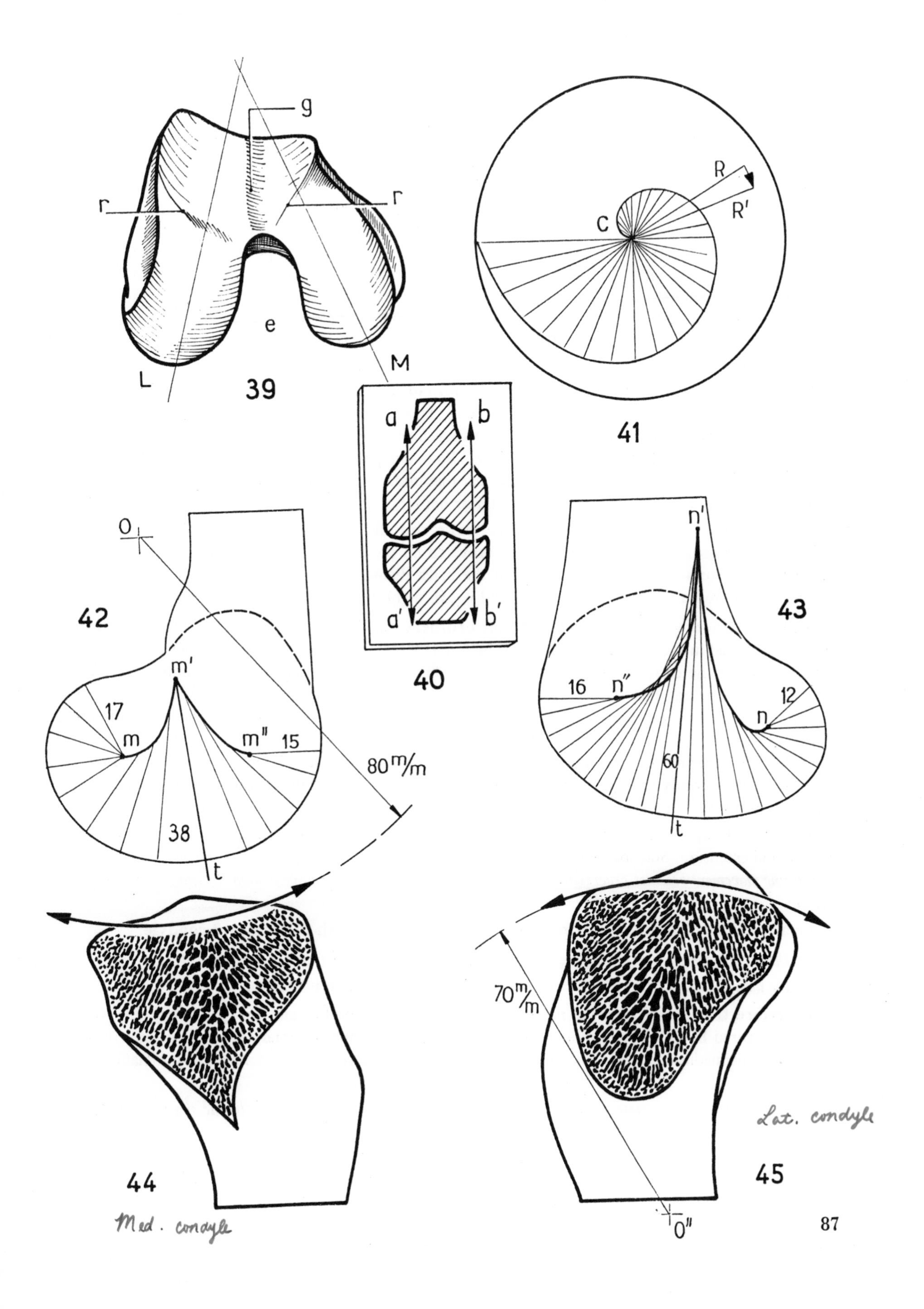
g
r
r
e
L
M
39
C
R
R'
41
a
b
a'
b'
40
O
42
m'
17
m
m''
15
80 m/m
38
t
n'
43
16
n''
12
n
60
t
70 m/m
Lat. condyle
44
45
Med. condyle
O''

THE MOVEMENTS OF THE FEMORAL CONDYLES ON THE TIBIAL CONDYLES DURING FLEXION AND EXTENSION

The rounded shape of the condyles could suggest that they roll over the tibial condyles, but this is wrong. In fact when **a wheel rolls on the ground without sliding** (fig. 46), *to each point on the ground corresponds a single point on the wheel* so that the distance covered on the ground (OO″) is exactly equal to the portion of the circumference which has rolled over the ground (distance between the point marked by a triangle and that marked by a rectangle). If this were so (fig. 47), after a certain measure of flexion (position II) the femoral condyle would tip over behind the tibial condyle—i.e. dislocation of the joint—or else the tibial condyle would need to be longer. The possibility of a simple rolling movement of the femoral condyle is precluded by the fact that *the length of the circumference of the femoral condyle is twice as great as the length of the tibial condyle.*

Let us now assume that **the wheel slides without rolling** (fig. 48): *therefore to one point on the ground corresponds a segment of the circumference of the wheel.* This is what happens when a car wheel 'spins' when starting on a frosty road. Such a type of sliding movement could conceivably explain the movements of the femoral condyle on the tibial condyle (fig. 49). So to one point on the tibial condyle would correspond all the points of the condylar surface. But it is clear that under these conditions *flexion would be prematurely checked by* the impact of the femur on the posterior border of the tibial condyle (arrow).

It is also possible to imagine that **the wheel rolls and slides simultaneously** (fig. 50): it skids but still moves forward. Therefore to the distance covered on the ground (OO′) corresponds a length much greater on the surface of the wheel (between the black diamond and the black triangle) than could be obtained by 'unrolling' the wheel on the ground between (black diamond and white triangle).

The experiment performed by the Weber brothers (fig. 51) in 1836 showed that the last mechanism actually operated in life. For various positions between extreme extension and extreme flexion they marked on the cartilage the corresponding points of contact between the femoral and tibial condyles. They then noted on the one hand that *the point of contact on the tibia moved backwards during flexion* (black triangle: extension, black diamond: flexion) and on the other hand that the distance between the points of contact marked on the femoral condyle was twice as long as that between the corresponding points on the tibial condyle. This experiment proves indisputably that the **femoral condyle rolls and slides simultaneously over the tibial condyle**. This is after all the only way that posterior dislocation of the femoral condyle can be avoided while a greater range of flexion remains possible (160°: compare flexion in figs. 49 and 51).

More recent experiments (Strasser 1917) have shown that the ratio of rolling to sliding varies during flexion and extension. Starting from full extension *the femoral condyle begins to roll without sliding and then the sliding movement becomes progressively more important so that at the end of flexion the condyle slides without rolling.*

Finally, the length over which pure rolling takes place **varies with the femoral condyle**:

for the medial condyle (fig. 52) pure rolling occurs only during the first 10° to 15° of flexion;

for the lateral condyle (fig. 53) this rolling goes on until 20° flexion is reached.

Therefore the lateral condyle rolls far more than the medial and this partly explains why the distance covered by the lateral femoral condyle over the corresponding tibial condyle is greater than that covered by the medial condyle. This important fact will be considered again in relation to automatic rotation (p. 134).

It is also interesting to note that the 15° to 20° of initial rolling corresponds to the normal range of the movements of flexion and extension during ordinary walking.

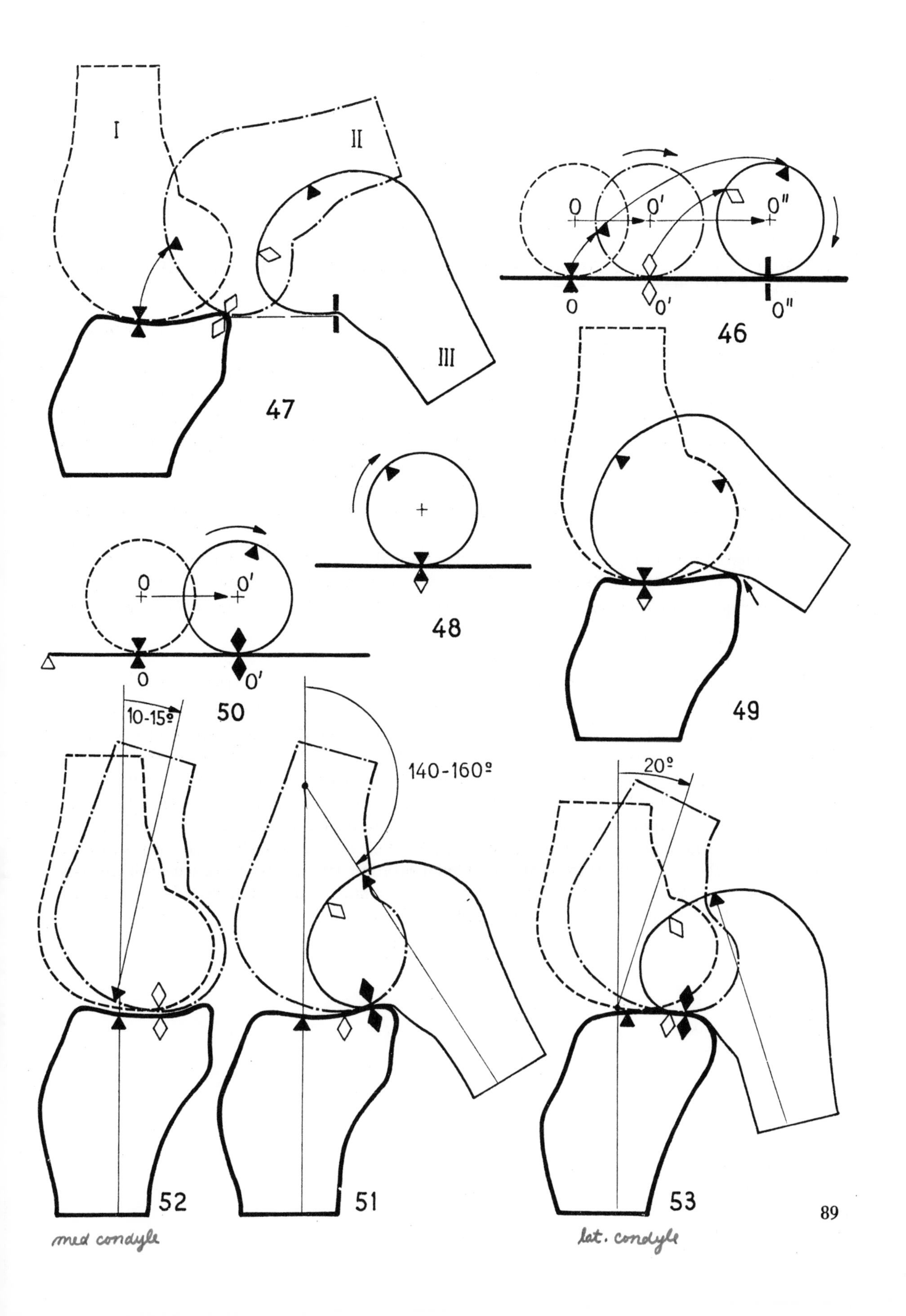
I
II
III
47
0
0'
0"
46
48
0
0'
50
49
10-15º
140-160º
20º
52
51
53
med condyle
lat. condyle

MOVEMENTS OF THE FEMORAL CONDYLES ON THE TIBIAL CONDYLES DURING AXIAL ROTATION

It will become apparent later why axial rotation can only take place when the knee is flexed. **In the neutral position for axial rotation** (fig. 54), with the knee flexed, the posterior part of the femoral condyles is in contact with *the middle part of the tibial condyles*. This is clearly illustrated by the diagram (fig. 55), where the outline of the femoral condyles (transparent) can be seen superimposed on the outline of the tibia condyles (striped). It is also clear from this diagram that with flexion of the knee the intercondylar tibial tubercles have moved clear of the intercondylar notch of the femur, where they normally lodge during extension (this is one of the reasons preventing axial rotation during extension).

During **lateral rotation** of the tibia on the femur (fig. 56) the lateral femoral condyle moves forward over the lateral tibial condyle while the medial femoral condyle moves backward over the medial tibial condyle (fig. 57).

During **medial rotation** (fig. 58) the converse takes place: the lateral femoral condyle moves posteriorly and the medial femoral condyle moves anteriorly over their respective tibial condyles (fig. 59).

The **anteroposterior movements of the femoral condyles** over their respective tibial condyles vary with the condyle:

the **medial condyle** (fig. 60) moves relatively little (l);

the **lateral condyle** (fig. 61), on the other hand, moves about twice as much (L) and in so doing it comes to lie slightly higher than the medial condyle. This difference in height is small but real (e).

The difference in the shape of the two tibial condyles is reflected in **the configuration of the intercondylar tubercles**. A horizontal section of these tubercules at level XX′ (fig. 62) shows that the lateral aspect of the lateral tubercle is *convex* anteroposteriorly (like the lateral tibial condyle) whereas the medial surface of the medial tubercle is *concave* (like the medial tibial condyle): the medial tubercle is also *clearly higher than the lateral tubercle*. As a result of these two features, the medial tubercle joints out as a stop-shoulder against which the medial femoral condyle knocks, while the lateral condyle easily moves past the corresponding tubercle. Therefore **the real axis of axial rotation** does not pass between the two tubercles but *through the medial tubercle*. This displacement of the 'centre' medially is reflected in the greater movement of the lateral condyle as shown previously.

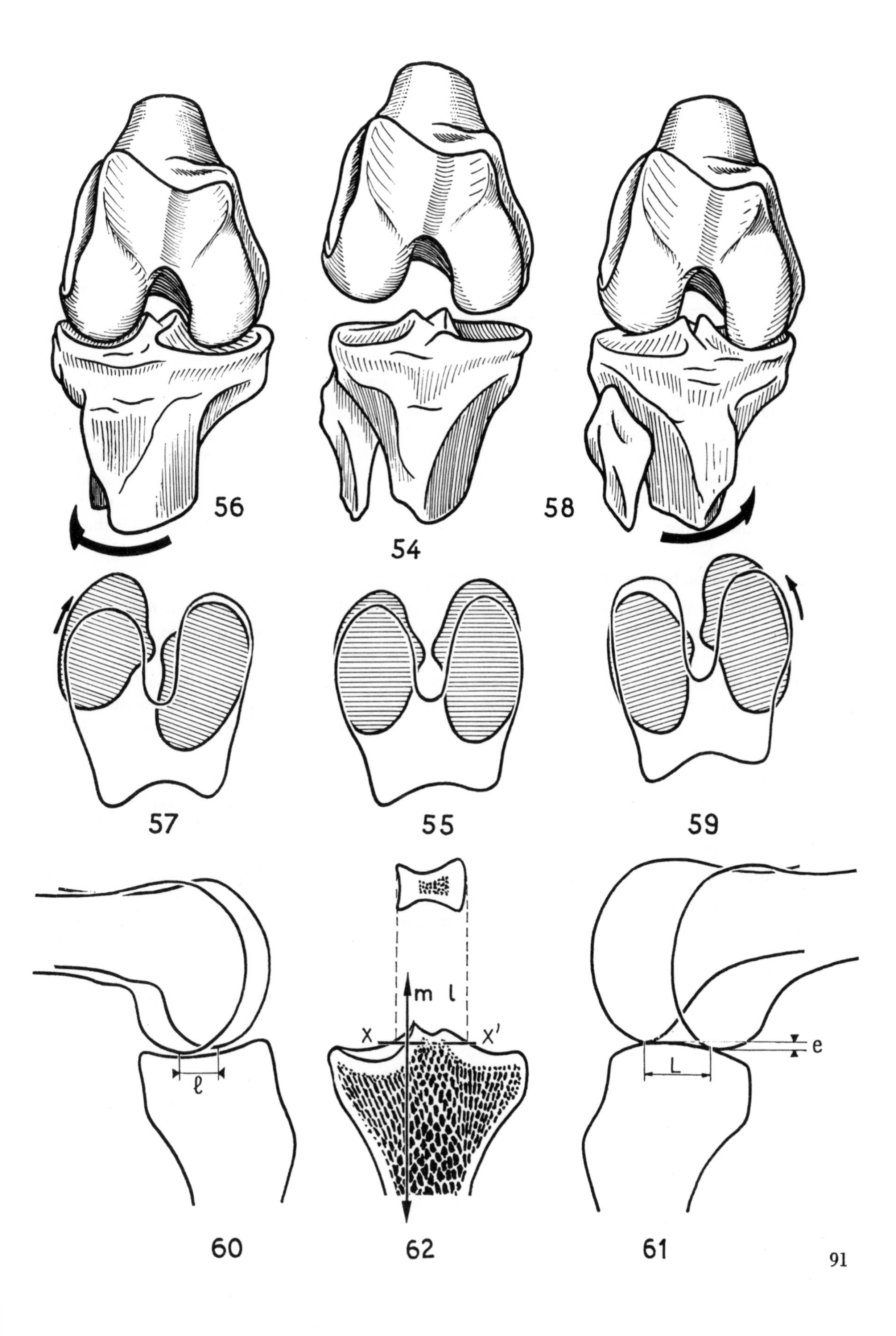
56
54
58
57
55
59
m
l
X
X'
ℓ
L
e
60
62
61

THE CAPSULE OF THE KNEE

The general form of the capsule (fig. 63) can be easily understood if it is compared to a *cylinder* which is invaginated posteriorly (as shown by the white arrow). This leads to the formation of a *partition* in the sagittal plane, which almost divides the cavity into a medial and a lateral half; its relations to the cruciate ligaments will be studied later (p. 116). On the anterior surface of the cylinder a *window* is cut to receive the patella. The upper and lower ends of the cylinder are attached to the femur and the tibia respectively.

The **attachment of the capsule to the tibia** is relatively simple (fig. 64, dots). It is attached to the anterior, medial and lateral aspects of the tibial condyles. Posteriorly *at the level of the popliteal notch* it is reflected forwards as a sleeve to be attached to the edges of the posterior intercondylar fossa along the margins of the condyles. It then runs *between the two intercondylar tubercles* and winds itself round the insertion of the anterior cruciate ligament in the middle of the anterior intercondylar area. Thus the tibial insertions of the anterior (1) and posterior (2) cruciate ligaments are *extracapsular* and lie outside the joint cavity.

The femoral attachment of the capsule is slightly more complex (figs. 65 to 68):

anteriorly (fig. 65), the capsule is attached to the bone along the edges of the shallow fossa overlying the patellar surface (7). At this point the capsule forms a deep recess (figs. 67 to 68), known as the *suprapatellar bursa* (5); its importance will emerge later (p. 102);

medially and laterally (figs. 65 to 68) it is inserted along the margins of the patellar surface forming the *parapatellar recesses* (p. 102) and then along the edges of the articular surfaces of the condyles like a *banister* (8); on the lateral condyle the capsular insertion lies above the insertion of the popliteus tendon (P), which is therefore *intracapsular* (see figs. 131 and 186);

posteriorly (fig. 66) the capsule is inserted round the posterosuperior border of the articular surfaces of the condyles, just distal to the origins of the two heads of the gastrocnemius (g). The capsule therefore lines the deep surface of these muscles and separates them from the condyles; in this area the capsule is thickened to form what can be called '*condylar plates*' (6) (p. 112);

in the intercondylar notch (figs. 67 to 68: the femur has been cut in the sagittal plane) the capsule is attached to the opposing surfaces of the condyles along the articular cartilage and then to the depths of the notch, which it bridges. Its insertion to the medial condyle (fig. 67) runs *below the femoral attachment of the posterior cruciate ligament* (4). Its insertion to the lateral condyle (3) lies *between the articular cartilage and the femoral attachment of the anterior cruciate ligament* (fig. 68).

Here too, therefore, the *insertion of the cruciate ligaments is extracapsular* and lies outside the joint cavity.

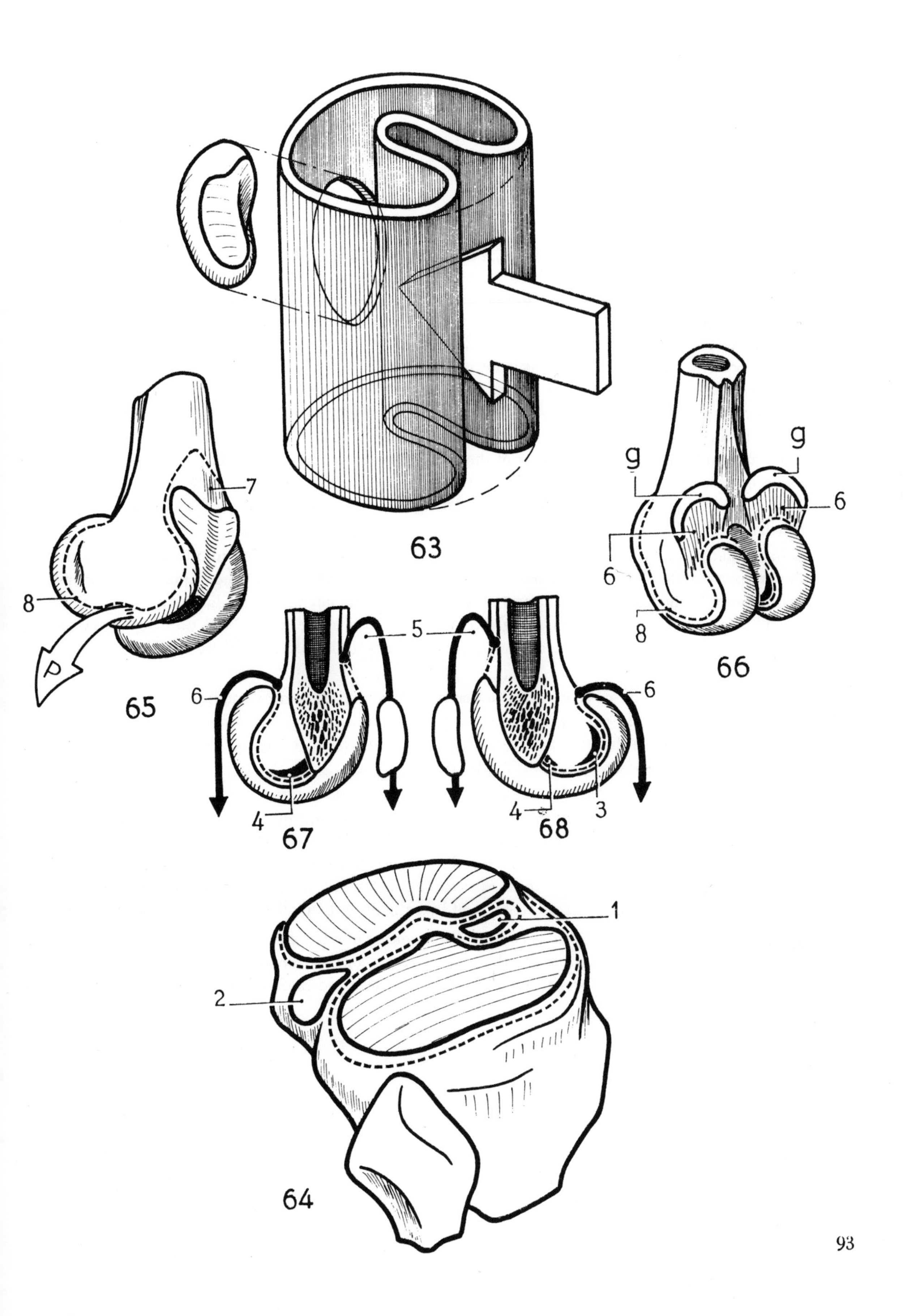
63
9
9
7
6
6
8
8
5
6
6
66
65
4
67
4
68
3
1
2
64

THE INFRAPATELLAR FOLD—THE CAPACITY OF THE JOINT

The empty space, bounded by the anterior intercondylar fossa of the tibia, the ligamentum patellae and the inferior aspect of the patellar surface of the femur, is filled by a considerable pad of adipose tissue known as **the infrapatellar pad** (fig. 69). This pad (1) has the shape of a *quadrangular pyramid* with its base resting on the posterior surface (2) of the ligamentum patellae (3) and overlapping the anterior part of the anterior intercondylar fossa. Its superior aspect (4) is strengthened by a fibro-adipose band attached to the apex of the patella and lying in the intercondylar notch (figs. 69 and 70): this band is *the infrapatellar fold* (5). At the sides (fig. 70: the knee has been opened anteriorly and the patella tilted) the infrapatellar pad extends superiorly along the inferior part of the sides of the patella in the form of two fringe-like folds of fibro-adipose tissue, i.e. *the alar folds* (6). The infrapatellar pad acts as a 'stopgap' in the anterior compartment of the joint. During flexion it is compressed by the ligamentum patellae and spreads out on either side of the patellar apex.

The infrapatellar pad is the vestige of the *median septum,* which divides the joint into two cavities in the embryo up to four months. In the adult, there is normally a gap (fig. 69) between the infrapatellar pad and the median septum formed by the capsule (arrow I). Therefore the lateral and medial halves of the joint cavity communicate with each other via this gap as well as via the space lying above the pad (arrow II) and deep to the patella. Occasionally the median septum persists in the adult and these only communicate above the infrapatellar pad.

The **capacity of the joint cavity** varies under normal and pathological conditions. An effusion—hydrarthrosis or haemarthrosis—increases this capacity substantially (fig. 71), provided it accumulates *gradually*; the fluid collects in the suprapatellar bursa (SP), in the parapatellar recesses and posteriorly in the gastrocnemius bursa (GB) deep to the 'condylar plates'.

The distribution of the fluid within the cavity varies according to the *position of the knee.* **In extension** (fig. 72) the gastrocnemius bursa is compressed by tension of the gastrocnemius and the fluid *moves anteriorly* where it collects into the suprapatellar bursa and the parapatellar recesses. **In flexion** (fig. 73) these become compressed by tension of the quadriceps and the fluid *moves posteriorly*. Between full flexion and full extension there is *a position of so-called maximal capacity* (fig. 71), where the fluid within the cavity is under least tension. This position of semi-flexion is assumed by patients with an effusion as it is the least painful.

Normally the amount of *synovial fluid* is *very small* amounting to a few cubic centimetres. However, the movements of flexion and extension ensure that the articular surfaces are constantly bathed by fresh synovial fluid and thus assist in the proper nutrition of the cartilage and especially in the lubrication of the surfaces in contact (Vol. IV).

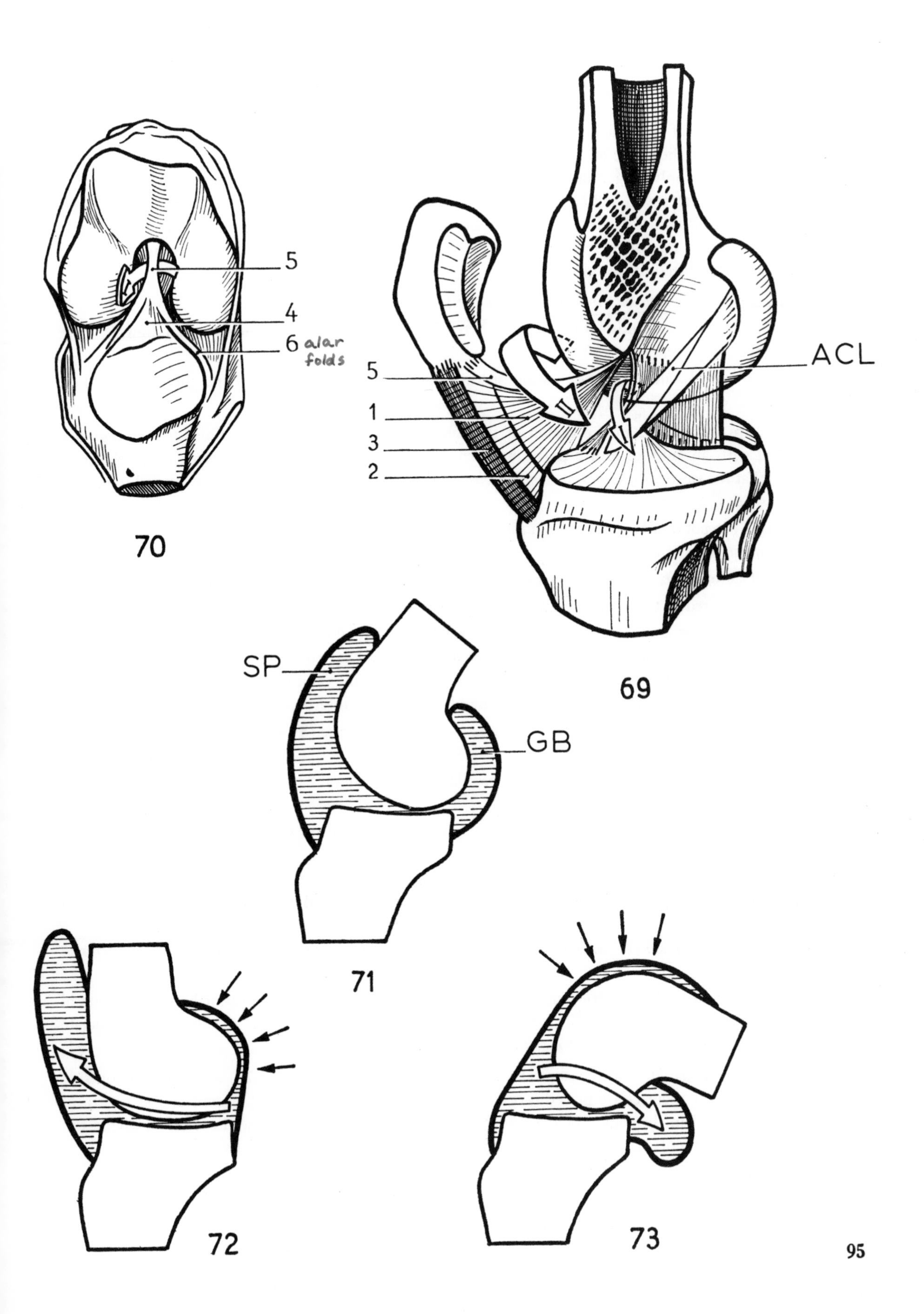
5
4
6 alar folds
70
ACL
5
1
3
2
69
SP
GB
71
72
73

THE MENISCI OR SEMILUNAR CARTILAGES OF THE KNEE

The lack of congruence of the articular surfaces (p. 86) is corrected by the interposition of **the menisci or semilunar cartilages**, the shape of which can easily be understood (fig. 74). If a sphere (S) is placed on a plane (P) contact occurs only tangentially with respect to the sphere. To increase the area of contact between the sphere and the plane, it will suffice to place between them a ring equal in volume to that bounded by the sphere, the plane P and the cylinder C which lies tangential to the sphere. Such a ring (shaded) has precisely the shape of a meniscus, *triangular in cross-section*, with the following **three surfaces** (fig. 75: the menisci have been lifted off the tibial condyles):

the superior surface (1), concave, in contact with the femoral condyles;

the peripheal surface (2), cylindrical in shape, adherent to the deep surface of the capsule (shown by the vertical stripes);

the inferior surface (3), almost plane, resting on the edges of the medial (MTC) and the lateral (LTC tibial condyles.

These rings are incomplete in the region of the intercondylar tubercles of the tibia so that they are *crescent-shaped* with an anterior and a posterior horn. The horns of the lateral meniscus come closer to each other so that *the meniscus is almost a complete circle* (in the shape of an O) whereas the medial meniscus is *a half-moon*, i.e. C-shaped.

These menisci are not unattached between the femoral and tibial surfaces and *have important attachments from the functional point of view*:

the deep surface of the *capsule* (fig. 76), as seen previously, is attached to the menisci;

each horn is anchored to the *tibial condyle* in the anterior and posterior intercondylar fossae respectively: the anterior horn of the lateral meniscus (4) just in front of the lateral intercondylar tubercle; the posterior horn of the lateral meniscus (5) just behind the lateral intercondylar tubercle; the posterior horn of the medial meniscus (7) in the posteromedial angle of the posterior intercondylar fossa; the medial horn of the medial meniscus (6) in the anteromedial angle of the anterior intercondylar fossa;

the two anterior horns are linked by the *transverse ligament of the knee* (8), which is itself attached to the patella by strands of the infrapatellar pad;

fibrous bands run from the lateral edges of the patella (P) to the lateral borders of each meniscus forming the *menisco-patellar fibres* (9);

the *medial collateral ligament* (MCL) of the knee is attached by its deep fibres to the internal border of the medial meniscus;

the *lateral collateral ligament* (LCL), however, is separated from its corresponding meniscus by the tendon of the popliteus (Pop) which sends a fibrous expansion (10) to the posterior border of the *lateral meniscus* (LM);

the *semimembranosus tendon* (11) also sends a fibrous expansion to the posterior edge of the *medial meniscus* (MM);

finally separate fibres of the posterior cruciate ligament are inserted into the posterior horn of the lateral meniscus forming the *menisco-femoral ligament* (12). There are also a few fibres of the anterior cruciate ligament inserted into the *anterior horn of the medial meniscus* (fig. 135).

The coronal section (fig. 76) and the sagittal sections, seen from the inside (fig. 77) and from the outside (fig. 78), show how the menisci are placed *between the tibial and femoral condyles*, except at the centre of each tibial condyle and in the region of the intercondylar tubercles, and also how the menisci divide the joint into two compartments: the suprameniscal and the inframeniscal compartments (fig. 76).

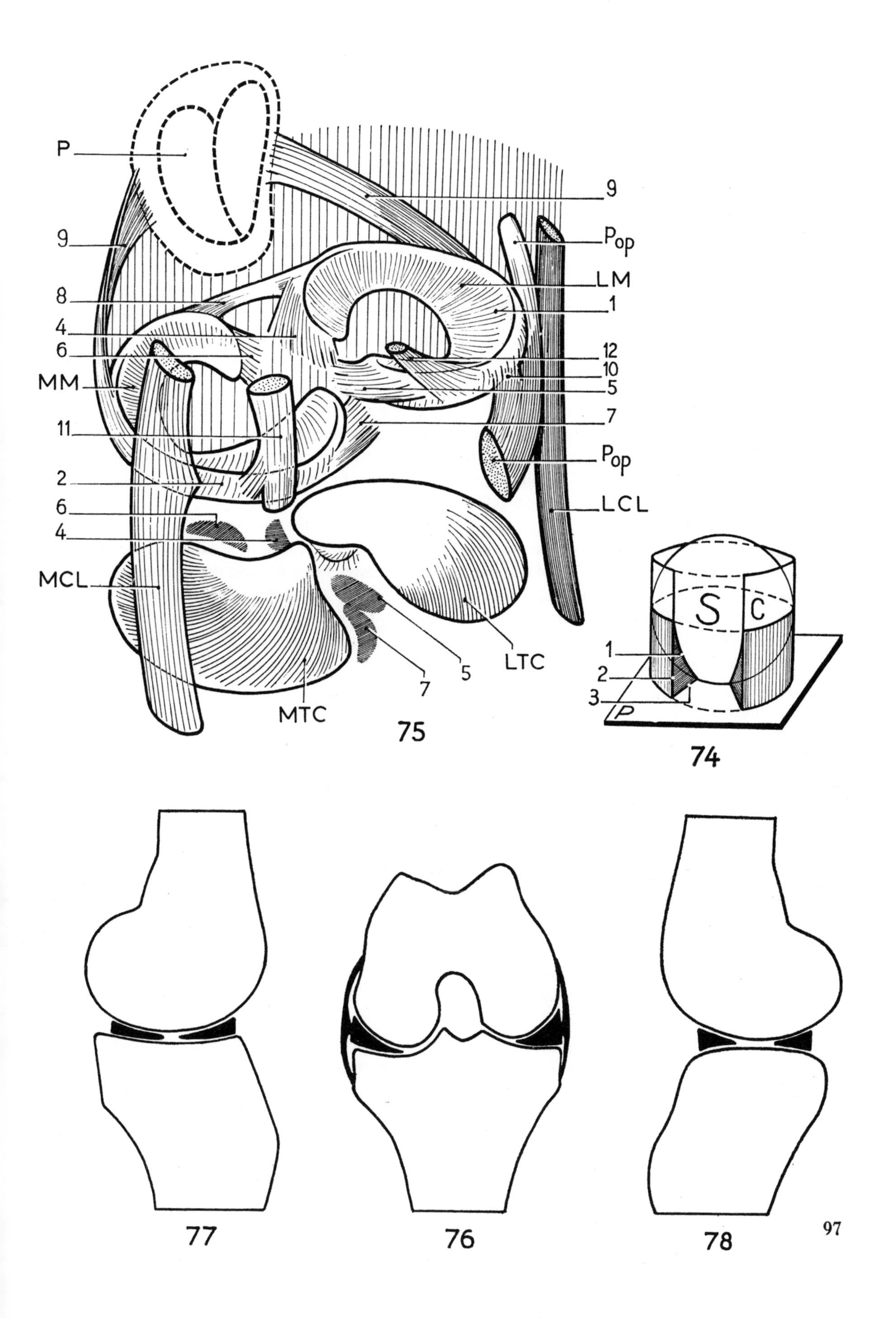

75

74

77

76

78

MOVEMENTS OF THE MENISCI DURING FLEXION AND EXTENSION

As shown before (p. 88), the point of contact between the femoral and tibial condyles moves posteriorly during flexion and anteriorly during extension; the menisci follow these movements, as is easily demonstrated in an anatomical preparation containing only the ligaments and the menisci. *In extension* (fig. 79), the *posterior part of the tibial condyle becomes exposed*, especially the lateral condyle (LTC). *In flexion* (fig. 80) *the menisci (medial and lateral) come to overlie the posterior part of the tibial condyles*, especially the lateral meniscus which reaches as far as the posterior border of the lateral condyle.

When the menisci are viewed from above it is obvious that starting from the position of extension (fig. 81), *the menisci move posteriorly unequally*; in flexion (fig. 82), the lateral meniscus (LM) has receded twice as far as the medial meniscus (MM). In fact, the posterior displacement of the medial meniscus is 6 mm., that of the lateral meniscus 12 mm.

It is also clear from these diagrams that, while they recede, *the menisci become distorted*. This is due to the fact that they have two fixed points, their anterior and posterior horns, while the rest of the structure is freely mobile. The lateral meniscus undergoes a greater degree of distortion and posterior displacement because its horns are attached much closer together.

The menisci play an important part as *an elastic coupling which transmits any compression forces* between the femur and the tibia (black arrows, figs. 84 and 85). It is worth noting that during extension the femoral condyles present their *greatest radii of curvature* to the tibial condyles (fig. 83) and the menisci are *tightly interposed* between the articular surfaces, These two factors *promote the transmission of compression forces* during full knee extension. On the other hand, during flexion, the femoral condyles display their shortest radii of curvature (fig. 86) and the menisci are only partially in contact with these condyles (fig. 88). These two factors, along with relaxation of the collateral ligaments (p. 106), *favour mobility at the expense of stability*.

After defining the movements of the menisci, the factors involved in these movements call for discussion. These factors fall into *two groups*: passive and active.

There is only **one passive element** involved in the displacement of the menisci: the *femoral condyles push the menisci anteriorly* just as a cherrystone is pushed forward between two fingers. This apparently over-simple mechanism is perfectly obvious when one studies an anatomical preparation where all the connections of the menisci have been severed except for the attachments of their horns (figs. 79 and 80). The surfaces are slippery and the 'wedge' of the meniscus is pushed anteriorly between the 'wheel' of the femoral condyle and the 'ground' of the tibial condyle (as a lock mechanism it is wholly inefficient).

The **active mechanisms** are numerous:

during extension (figs. 84 and 85), the menisci are pulled forward by the *meniscopatellar fibres* (1), which are stretched by the anterior movement of the patella (p. 104) and this draws the transverse ligament forward. In addition, the posterior horn of the lateral meniscus is pulled anteriorly (fig. 85) by the tension developed in the *meniscofemoral ligament* (2), as the posterior cruciate ligament becomes taut (p. 122);

during flexion, the medial meniscus (fig. 87) is drawn posteriorly by *the semimembranosus expansion* which is attached to its posterior edge, while the anterior horn is pulled anteriorly by the *fibres of the anterior cruciate ligament* attached to it (4); the *lateral meniscus* (fig. 88) is drawn posteriorly by the popliteus expansion (5).

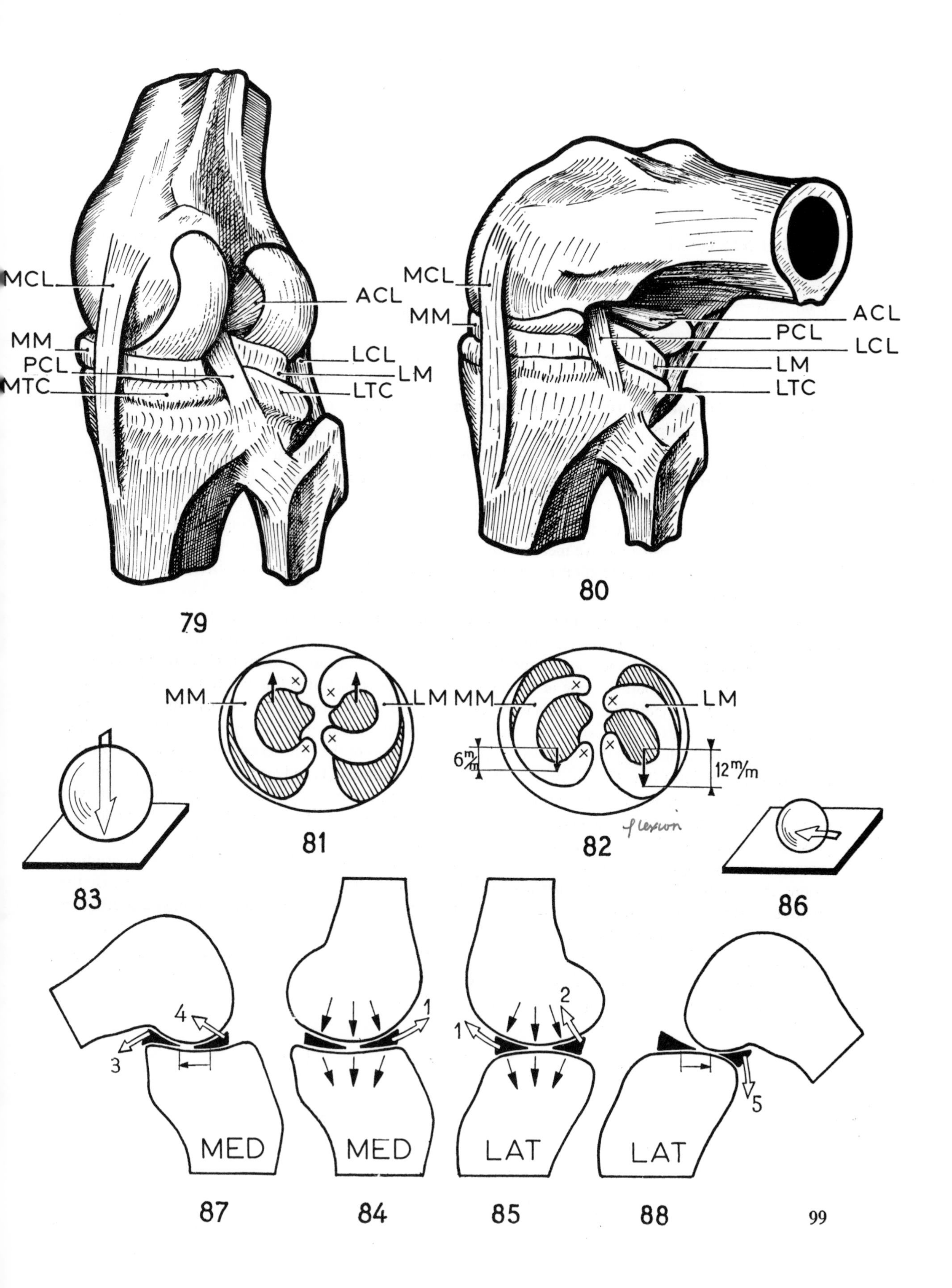

MCL
MM
PCL
MTC
ACL
LCL
LM
LTC
79
MCL
MM
ACL
PCL
LCL
LM
LTC
80
MM
LM
MM
LM
6m/m
12m/m
81
82
flexion
83
86
4
3
1
2
1
5
MED
MED
LAT
LAT
87
84
85
88

MOVEMENTS OF THE MENISCI DURING AXIAL ROTATION; LESIONS OF THE MENISCI

During the movements of axial rotation, the menisci follow *exactly* the displacements of femoral condyles (p. 90). Starting from the neutral position (fig. 89) they can be seen to move on the tibial condyles in the opposite direction:

during lateral rotation (fig. 90), the lateral meniscus (LM) is pulled towards the anterior part (1) of the tibial condyle while the medial meniscus (MM) is drawn posteriorly (2);

during medial rotation (fig. 91), the medial meniscus (MM) moves forward (3) while the latelar meniscus (LM) recedes (4).

Here again the menisci during their movements *become distorted* about their fixed points, i.e. the attachments of their horns. The total range of the movement of the lateral meniscus (1+4) is twice as great as that of the medial meniscus (2+3).

These displacements of the menisci during axial rotation are mostly passive—being drawn by the femoral condyles—but there is also an active mechanism involved. The *meniscopatellar fibres become taut* as a result of movement of the patella in relation to the tibia (p. 104) and this tension in these fibres draws one of the menisci anteriorly.

During movements of the knee **the menisci can be injured** if they fail to follow the movements of the femoral condyles on the tibial condyles; they are thus 'caught unawares' in an abnormal position and are 'squashed between the anvil and the hammer'. This happens, for instance, **during violent extension of the knee** (i.e. kicking a football): one of the menisci fails to move forwards (fig. 92) and is caught between the femoral and tibial condyles as the tibia is forcefully applied to the femur. This mechanism, very common among footballers, leads to *transverse tears* (fig. 97, a) or to *detachment of the anterior horn* (b), which then becomes folded on itself. The other mechanism producing lesions of the menisci involves a **twisting movement of the knee joint** (fig. 93), which combines *lateral displacement* (1) and *lateral rotation* (2). The medial meniscus is then pulled towards the centre of the joint under the convexity of the medial femoral condyle; when the joint is extended it is caught off guard and crushed between the two condyles with the following possible consequences: (a) a *longitudinal splitting of the meniscus* (fig. 94) or (b) a *complete detachment of the meniscus from the capsule* (fig. 95) or (c) a *complex tear of the meniscus* (fig. 96). In all these longitudinal lesions the central free part of the meniscus can rear itself up into the intercondylar notch so that the meniscus assumes the shape of a *bucket-handle*. This type of lesion is very common among footballers (during falls on a flexed leg) and among miners who have to work crouched in narrow seams of coal.

As soon as a meniscus is torn, the injured part fails to follow the normal movements and becomes wedged between the femoral and tibial condyles. The knee as a result '*locks*' in a position of flexion, which is more marked the more posterior the rupture; *full extension is then impossible.*

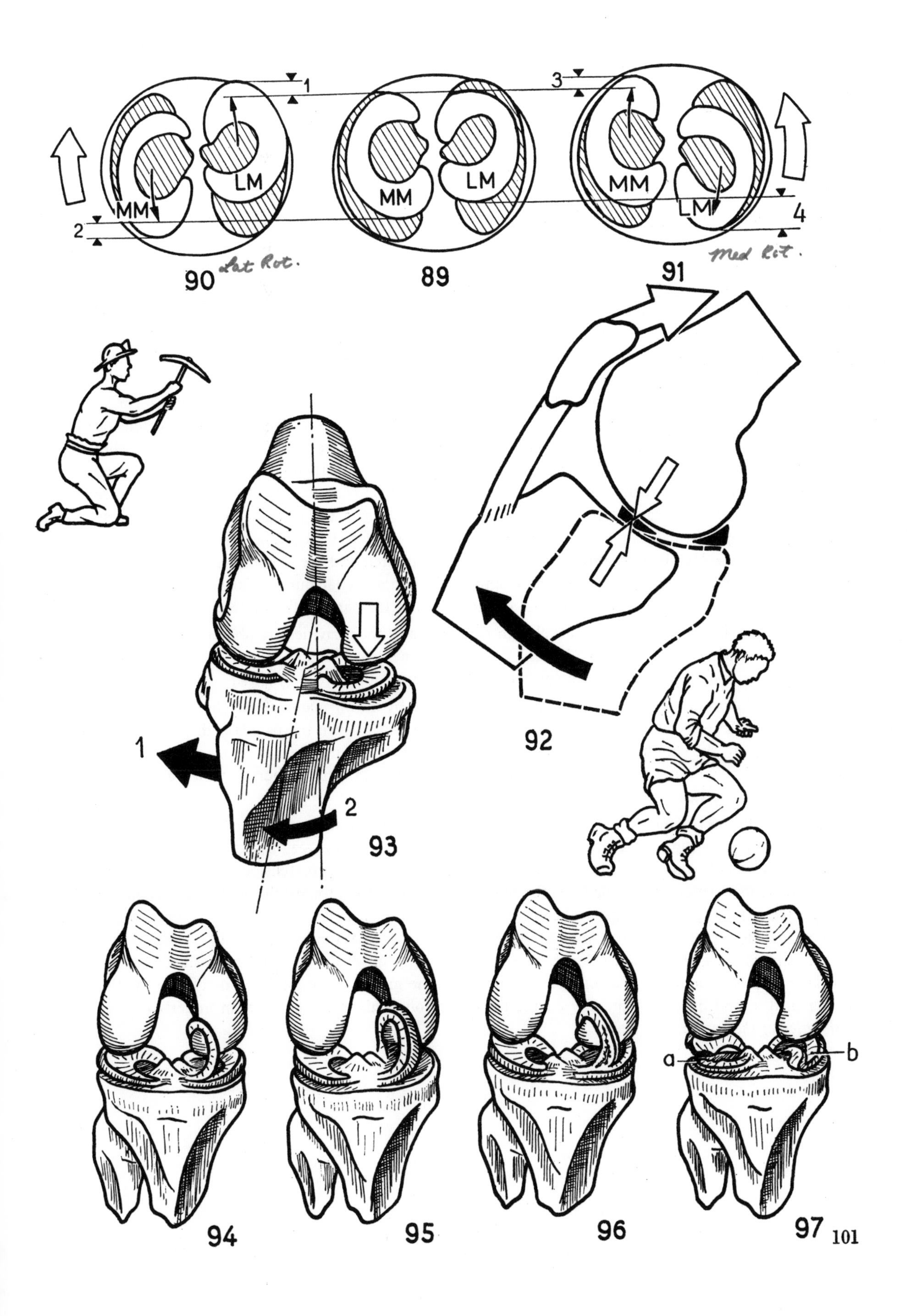

1
2
3
4
MM
LM
MM
LM
MM
LM
90
89
91
92
1
2
93
a
b
94
95
96
97

MOVEMENTS OF THE PATELLA ON THE FEMUR

The extensor apparatus of the knee slides on the lower end of the femur like **a cable on a pulley** (fig. 98, a). The patellar surface of the femur and the intercondylar notch (fig. 99) effectively form a deep vertical gutter (fig. 98, b), in the depths of which slides the patella. Thus the force of the quadriceps, directed obliquely superiorly and *slightly laterally*, is turned into a *strictly vertical force.*

The normal movement of the patella on the femur during flexion is therefore a vertical displacement along the central groove of the femoral patellar surface down to the intercondylar notch (fig. 100: based on X-ray studies). Thus the patella moves downwards *over a distance equal to twice its length* (8 cm.), while turning on itself about a transverse axis. Its deep surface, which looks directly posteriorly in extension (A), faces superiorly when the patella, at the end of its downward displacement in full flexion (B), comes to lie against the femoral condyles. This movement can thus be called '**circumferential displacement**'.

This important displacement is only possible when the patella is attached to the femur by connections of sufficient length. The capsule forms three recesses in relation to the patella (fig. 100) superiorly, the *suprapatellar bursa* (SP) and on either side the *parapatellar recesses* (PPR). When the patella slides under the condyles from A to B these three recesses become unpleated and the distance XX′ can become XX″ (i.e. four times greater) only because of the length of the suprapatellar bursa. Likewise, the distance YY′ can become YY″ (i.e. twice as great) because of the length of the parapatellar recesses.

When the inflammatory adhesions develop in these recesses their cavities are obliterated and *the patella is tightly held against the femur* (i.e. XX′ and YY′ became inextensible) and cannot slide down the central groove. This is one of the causes of the post-traumatic or post-infective '*stiff knee*'.

During its downward displacement the patella is followed by the *infrapatellar pad* (fig. 101) which moves from position ZZ′ to position ZZ″, i.e. through an angle of 180°. When the patella is displaced superiorly during extension, the suprapatellar bursa would be caught between patella and femur, were it not drawn upwards by the *articularis genu muscle* (AGM) (tensor of the suprapatellar bursa), which arises from the deep surface of the vastus intermedius.

Normally the patella moves only in the vertical plane and not transversely. It is in fact very strongly applied to its groove (fig. 102) by the quadriceps, the more so as the degree of flexion increases (a). At the end of extension (b) this appositional force is diminished and in hyperextension (c) it even tends to be reversed, i.e. to separate the patella from the femur. At this point (d) the patella tends *to be driven laterally* because the quadriceps tendon and the ligamentum patellae form *an angle obtuse laterally*. Lateral dislocation of the patella is prevented by the lateral lip of the patellar surface of the femur (fig. 103) *which is distinctly more prominent than the medial lip* (difference = e). If, as a result of a congenital malformation (fig. 104), this lateral lip is underdeveloped (i.e. is as prominent as the medial lip or less prominent) the patella is no longer held in and dislocates laterally during full extension. This is the mechanism underlying *recurrent dislocation of the patella.*

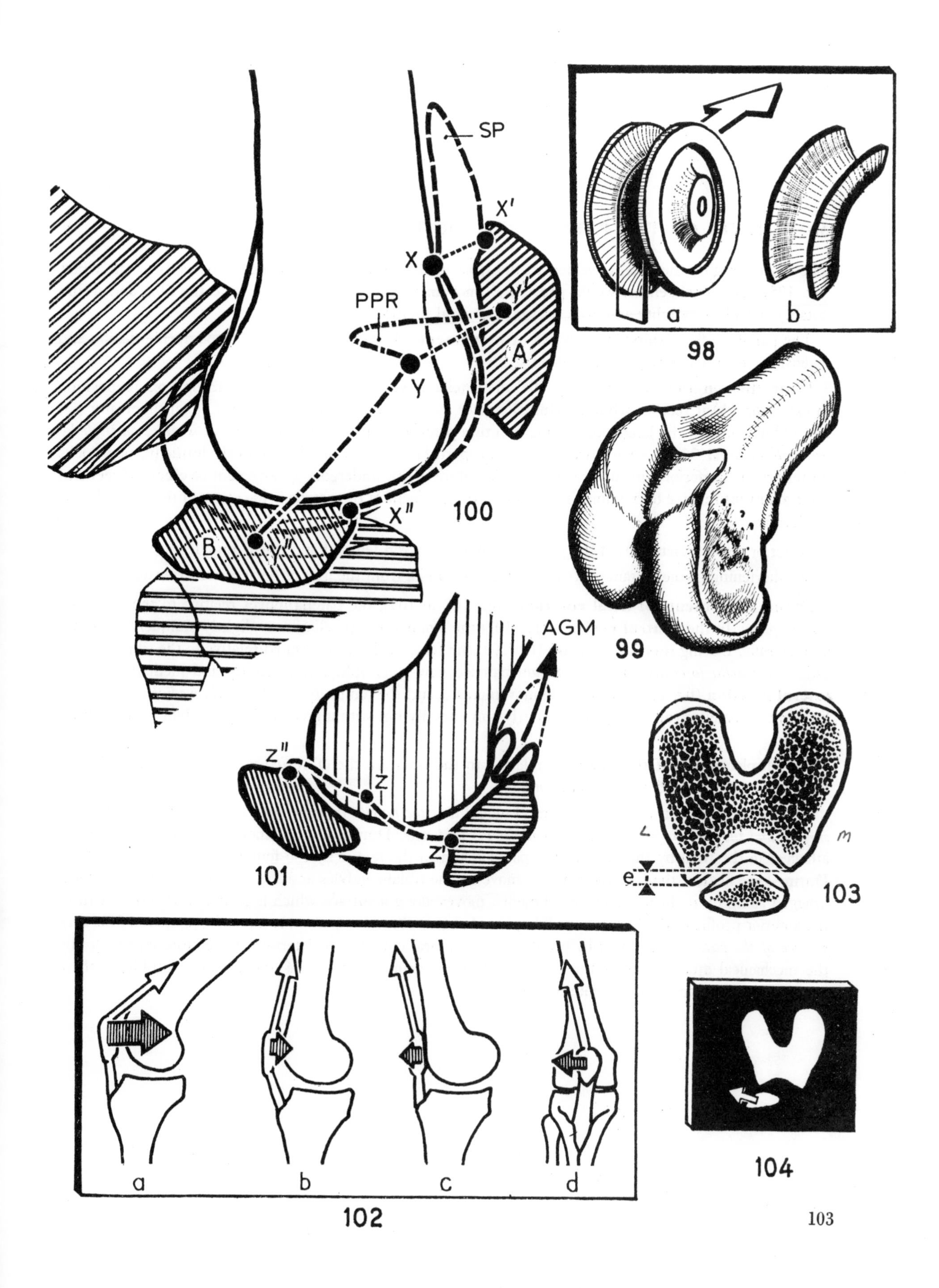
SP
X'
X
Y'
PPR
A
Y
X''
100
B
Y''
a
b
98
99
AGM
Z''
Z
Z'
101
e
103
a
b
c
d
102
104

MOVEMENTS OF THE PATELLA ON THE TIBIA

One could imagine the patella welded to the tibia in the form of an olecranon process as at the elbow (fig. 105). This arrangement would prevent all movements of the patella relative to the tibia and would notably curtail its mobility and even prevent any axial rotation.

The patella in fact exhibits **two types of movement** relative to the tibia, one type during flexion and extension, the other during axial rotation.

During **flexion and extension** (fig. 106) the patella moves in a sagittal plane. Starting from its position in extension (A) it recedes while moving along *the arc of a circle* with centre at the tibial tuberosity (O) and radius equal to the length of the pigamentum patellae. During this movement *it is tilted on itself by an angle of about* 35° in such a way that its deep surface, which faced posteriorly intitially, now looks posteriorly and inferiorly in full flexion (B). Therefore it also undergoes a movement of **circumferential displacement** relative to the tibia. This backward movement of the patella is the result of the *two following mechanisms*:

posterior displacement (D) of the point of contact between femoral and tibial condyles;

the shortening of the distance (R) separating the patella from the axis of flexion and extension (+).

During **movements of axial rotation** (figs. 107 to 109), the patella moves relative to the tibia in a *frontal plane*. In the **neutral position** (fig. 107) the ligamentum patellae runs slightly obliquely inferiorly and laterally. During **medial rotation** (fig. 108) the femur is laterally rotated relative to the tibia and this *drags the patella laterally*; the ligamentum patellae now runs obliquely inferiorly and medially. During **lateral rotation** (fig. 109) the opposite movements occur: the femur *draws the patella medially* so that the ligamentum patellae runs obliquely inferiorly and laterally but with a greater obliquity than in the neutral position.

The displacements of the patella in relation to the tibia are therefore *indispensable* for movements of *flexion and extension* and of *axial rotation.*

It is easy to demonstrate on a mechanical model that *the patella is responsible for the shape of the patellar surface and the anterior aspects of the condyles of the femur*. During its movements the patella is in effect attached to the tibia by the ligamentum patellae and to the femur by the femoropatellar fibres (p. 106). During knee flexion, the femoral condyles move on the tibial condyles and the deep surface of the patella, dragged along by its ligamentous attachments, moves along a surface which is *geometrically* equivalent to the anterior profile of the femoral condyles, i.e. *the curve that encompasses the successive positions of the deep surface of the patella.* The anterior profile of the femoral condyles is determined therefore essentially by the mechanical attachments of the patella and their arrangement just as the posterior profile of these condyles depends upon the cruciate ligaments (p. 120).

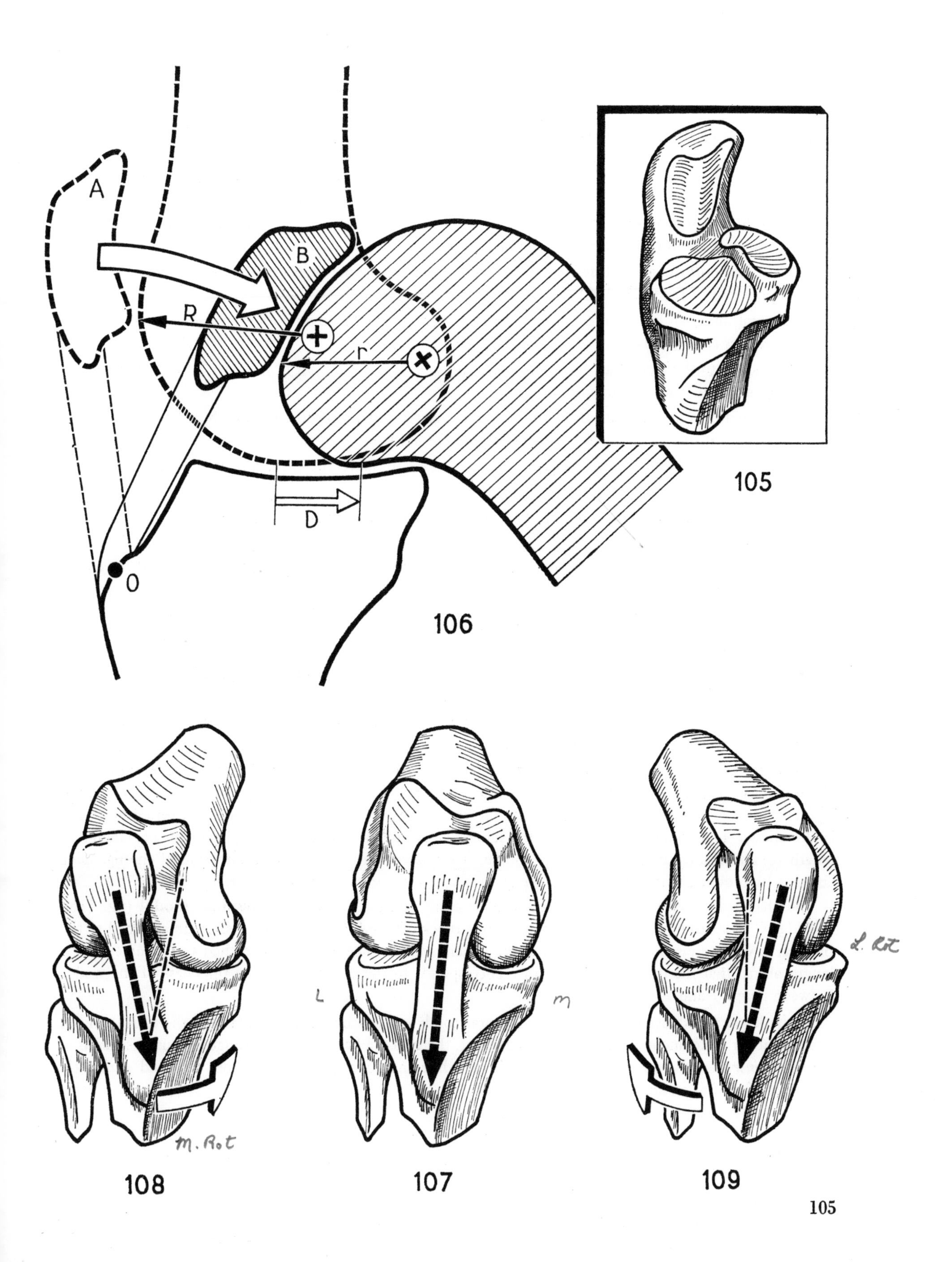
A
B
R
r
D
O
105
106
L
m
M. Rot
L. Rot
108
107
109

THE COLLATERAL LIGAMENTS OF THE KNEE

The stability of the knee depends upon the action of *powerful ligaments*, i.e. the cruciate and the collateral ligaments.

The **collateral ligaments** strengthen the articular capsule on its medial and lateral aspects. They are therefore responsible for the transverse stability of the knee during extension.

The medial collateral ligament (fig. 110) runs from the medial aspect of the medial femoral condyle to the upper end of the tibia (MCL):

its femoral insertion lies on the postero-superior aspect of the condyle, *posterior and superior to the line of the centres of curvature* (xx′) *of the condyle* (p. 86);

it is attached to the medial aspect of the tibia *posterior to the insertions of the three medial tibial muscles* (sartorius, gracilis and semimembranosus);

its anterior fibres are separate from those of the capsule but its posterior fibres blend intimately with those of the capsule *at the medial border of the medial meniscus;*

it runs obliquely inferiorly and anteriorly, i.e. it crosses in space the direction of the lateral collateral ligament (arrow A).

The lateral collateral ligament (fig. 111) runs from the outer surface of the lateral condyle to the head of the fibula (LCL).

it is attached to the femur *superior and posterior to the line of the centres of curvature* (yy′) *of the lateral condyle;*

it is attached to the *fibular head* anterior to its styloid process and deep to the insertion of the biceps;

it is *free of the capsule* along its entire course;

it *runs obliquely inferiorly and posteriorly* and so crosses in space the direction of the medial collateral ligament (arrow B).

In diagrams (figs. 110 and 111) the *meniscopatellar fibres* are seen (1 and 2) as well as the *alar folds* (3 and 4), which keep the patella against the femur.

The **collateral ligaments become taut during extension** (figs. 112 and 114) and **slackened during flexion** (figs. 113 and 115). Figures 112 and 113 show the difference (d) in length of the medial collateral ligament in the positions of extension and flexion; the change in the obliquity of its postero-inferior course is also shown. Figures 114 and 115 show the same changes in the lateral collateral ligament, i.e. change in length (e) and change in obliquity of its course; from extension to flexion, the direction of the ligament changes from oblique *inferiorly and posteriorly to oblique inferiorly and slightly anteriorly.*

The change in tension in the ligaments can be easily illustrated by **a mechanical model** (fig. 116). A wedge C slides from position 1 to position 2 on a block of wood B and fits into a strap ab attached to the block at a. When c moves from 1 to 2 the strap, which is taken to be elastic, is stretched and assumes a new length ab′ so that the difference in length e corresponds to the difference in the thickness of the wedge between points 1 and 2.

In the knee, as extension proceeds, the femoral condyle slides like a wedge between the tibial condyle and the upper attachment of the collateral ligament: the condyle behaves like a wedge because its *radius of curvature increases progressively postero-anteriorly*, while the collateral ligaments are attached *within the concavity of the line joining its centres of curvature.*

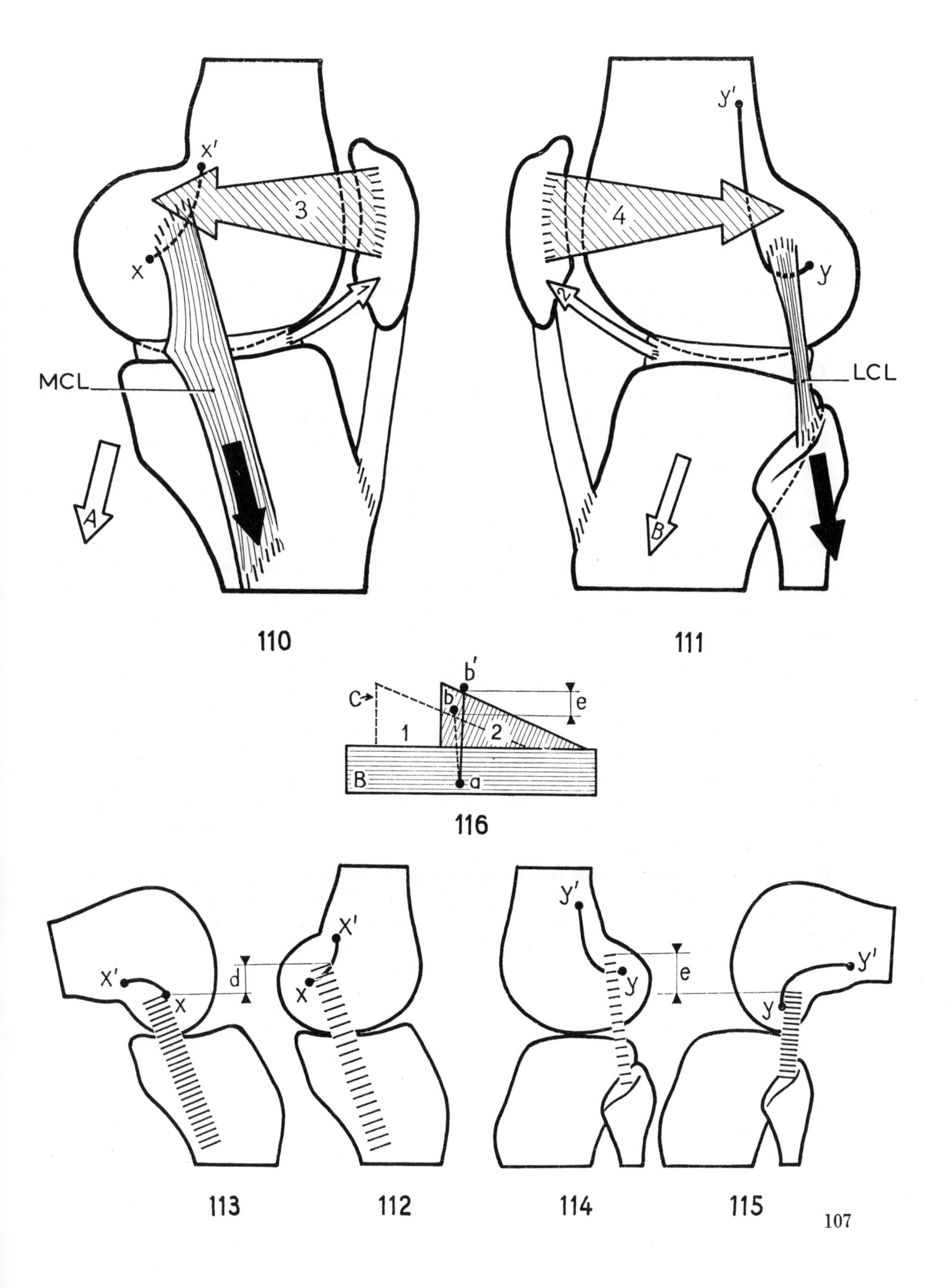
x'
3
x
MCL
A
y'
4
y
2
LCL
B
110
111
b'
c
b
e
1
2
B
a
116
x'
x
d
X'
x
y'
y
e
y'
y
113
112
114
115

THE TRANSVERSE STABILITY OF THE KNEE

The knee is subjected to considerable side-to-side stresses which are reflected in *the structure of the bony extremities* (fig. 117). As in the upper end of the femur, various bony trabeculae are present along the *lines of the mechanical stresses*:

The distal end of the femur contains *two main sets of trabeculae*: the one runs from the *cortical bone of the inner surface* of the femur and fans out into the ipsilateral condyle (compression forces) and into the contralateral condyle (traction forces); the other runs from *the cortical bone of the lateral surface* and fans out in a corresponding fashion. Another system of trabeculae runs horizontally to unite the two condyles.

The proximal end of the tibia has a similar set of trabeculae which start from the cortical bone of the medial and lateral surfaces and radiate out respectively into the ipsilateral condyle (compression forces) and into the contralateral condyle (traction forces). These two condyles are united by horizontal trabeculae.

Since the femoral axis runs inferiorly and *medially*, the force (F) applied to the superior aspect of the tibia is not strictly vertical (fig. 118) and can therefore be resolved into a vertical component (v) and a transverse component (t), which points horizontally and medially. This component (t) tends to tilt the joint medially and *exaggerate the physiological valgus* by widening the interspace medially (a). The medial ligaments normally prevent such a dislocation.

This transverse component (t) is greater the more marked the physiological valgus (fig. 119). For a force F_2 corresponding to a state of genu valgum (angle of valgus = 160°) the transverse component (t) is twice as great as it is with a normal degree of valgus (170°: F_1 and t_1). Hence *the more marked the genu valgum, the greater its tendency to increase in severity and the greater the demands imposed on the medial ligaments.*

During violent injuries involving the medial and lateral aspects of the knee the upper end of the tibia can be fractured. If the **force is applied to the medial aspect of the knee** (fig. 120), it tends to *straighten out the physiological valgus* and produces first a *fracture-dislocation of the medial tibial condyle* (1) and, if the force is strong enough, a *rupture of the lateral collateral ligament* (2). If the ligament snaps straight away the tibia escapes fracture.

When the **force is applied to the lateral aspect of the knee** (fig. 121), e.g. by the bumper of a car, the lateral femoral condyle is first of all slightly displaced medially and then becomes impacted into the lateral tibial condyle and eventually splits the cortical bone of the lateral aspect of the tibial condyle. This produces a *mixed type of fracture* of the lateral tibial condyle (i.e. impaction-dislocation).

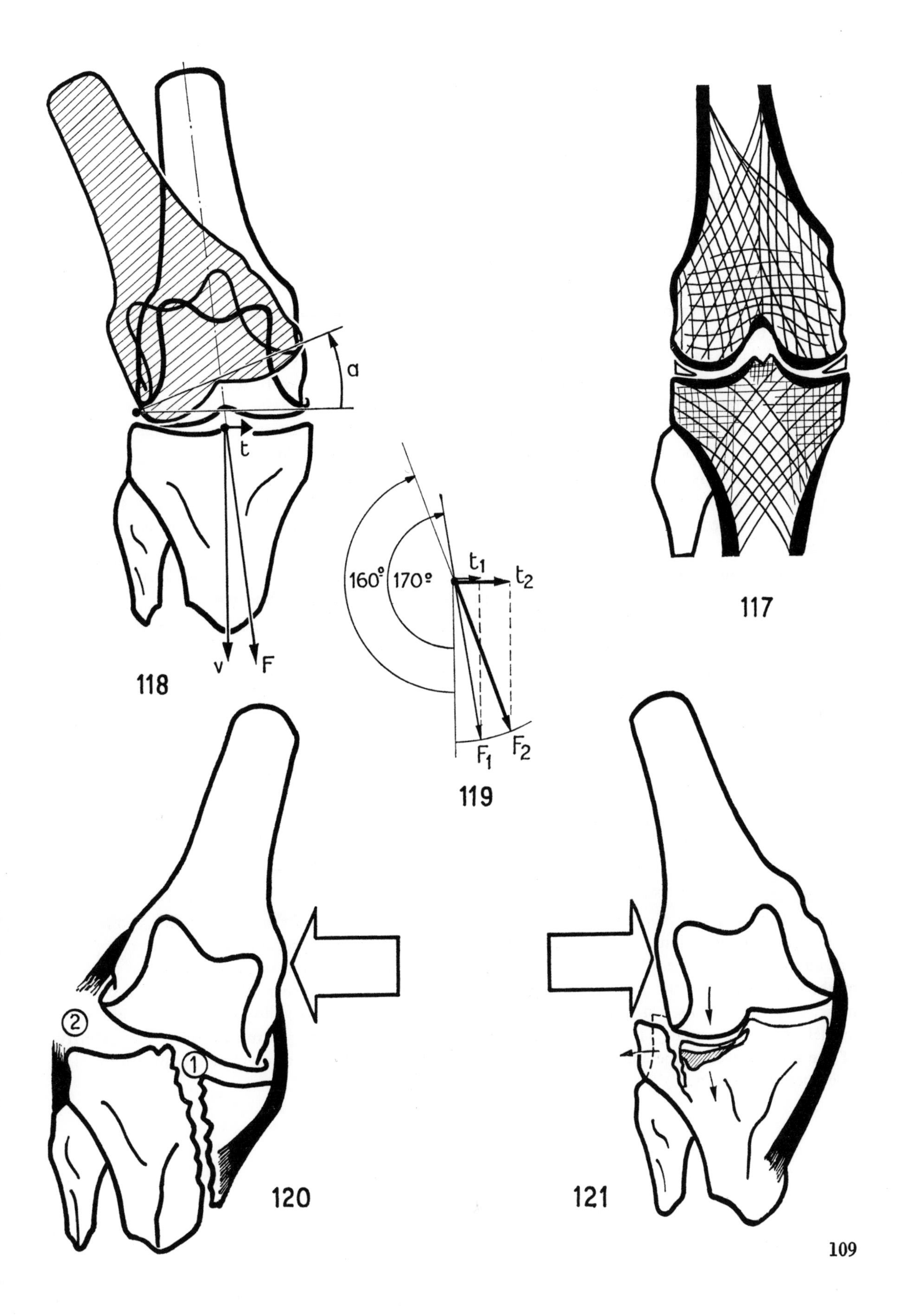
a
t
v
F
118
160°
170°
t_1
t_2
F_1
F_2
119
117
②
①
120
121

THE TRANSVERSE STABILITY OF THE KNEE—*(Continued)*

During walking and running the knee is continually subjected to side-to-side stresses. In certain postures the body is in a **state of imbalance, being tilted medially** relative to the supporting knee (fig. 122), this *tends to exaggerate the physiological valgus* and to open out the interspace of the joint *medially*. If the stress is too severe *the medial collateral ligament is torn* (fig. 123) leading to a **severe sprain of the medial collateral ligament** (it must be stressed that this severe degree of sprain does not result purely from the state of imbalance and requires in addition the application of a strong force).

When the body is in the **other state of imbalance, i.e. tilted laterally** relative to the supporting knee (fig. 124) this *tends to straighten out the physiological valgus* and to open out the interspace of the joint *laterally*. If a violent force is applied to the medial aspect of the knee the lateral collateral ligament can be torn as a result, i.e. the **severe sprain of the lateral collateral ligament** (fig. 125).

When **the knee is severely sprained,** *abnormal side-to-side movements can be demonstrated* occurring about an anteroposterior axis (figs. 1 and 2). For their demonstration the knee must be fully extended or even hyperextended (when normally the maximally stretched collateral ligaments prevent any sideways movements).

Therefore, if in hyperextension the leg can be displaced laterally (fig. 123), an *abnormal lateral movement* is said to be present and this indicates **tearing of the medial collateral ligament.**

If, on the other hand, the leg can be moved medially (fig. 125), an *abnormal medial movement* is present, indicating a **torn lateral collateral ligament.**

A severe sprain of the ligaments impairs the stability of the knee. In fact, when the lateral collateral ligament is torn the knee cannot resist the lateral stresses to which it is continually subjected (figs. 122 and 124).

When violent side-to-side stresses are applied during walking or running the collateral ligaments are not the only structures available to stabilise the knee; they are assisted by the *muscles* which form *true active ligaments* and play a vital part in securing the stability of the joint (fig. 126).

The lateral collateral ligament (LCL) is powerfully assisted by the *iliotibial tract* (1), which is tightened by the *tensor fasciae latae* whose contraction is shown in fig. 124.

The medial collateral ligament (MCL) is likewise assisted by contraction of *the medial tibial muscles* i.e. sartorius (2), semitendinosus (3), gracilis (4) whose contraction is shown in action in fig. 122.

The collateral ligaments are therefore 'lined' by thick muscle tendons. They are also assisted as powerfully by the **quadriceps** with its *straight* (S) and *cruciate* (C) fibres forming a predominantly fibrous canopy for the anterior aspect of the joint. The straight fibres prevent the opening out of the interspace of the joint on the same side, while the cruciate fibres prevent its opening out on the other side. Therefore each vastus muscle, by virtue of its two types of fibres of insertion, influences the stability of the knee both on its medial and lateral aspects. This highlights the significance of the quadriceps in **maintaining the stability of the knee** as well as the abnormal positions of the knee resulting from atrophy of the quadriceps (i.e. 'the knee that gives way').

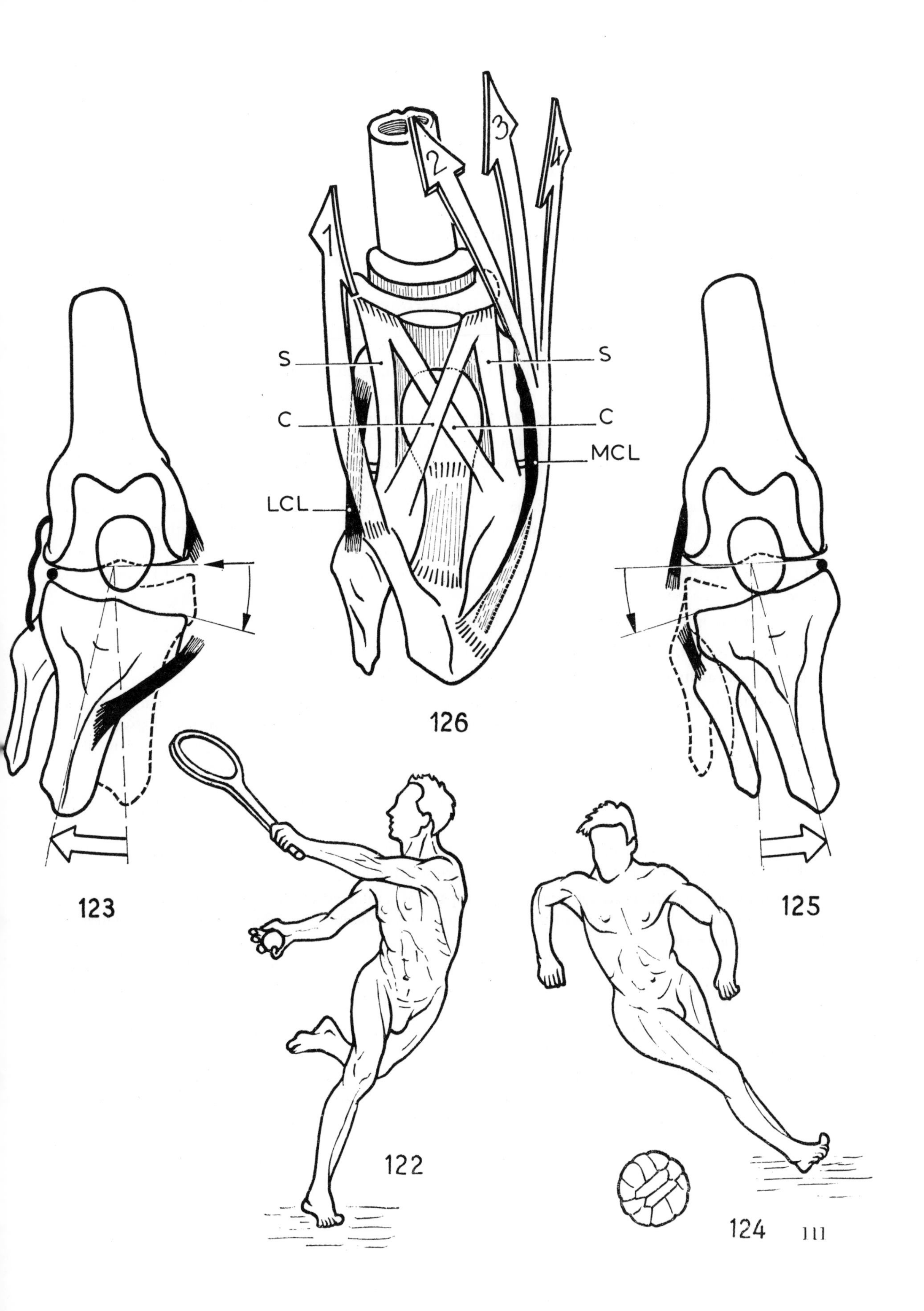
1
2
3
4
S
S
C
C
MCL
LCL
126
123
125
122
124

ANTEROPOSTERIOR STABILITY OF THE KNEE

The mechanism of stabilisation of the joint varies according to whether it is slightly flexed or hyperextended.

When **the knee is straight and very slightly flexed** (fig. 127), the force exerted by the body weight acts *behind the flexion and extension axis* of the knee and so the knee tends to flex further unless prevented by contraction of the quadriceps. Therefore in this position the *quadriceps is essential for the maintenance of the erect posture*. On the other hand, if the knee is **hyperextended** (fig. 128) the natural tendency for this hyperextension to increase is soon checked by the capsule and the related ligaments posteriorly (shown in black): *so this erect posture can be maintained without the quadriceps*. This explains why during paralysis of the quadriceps, the state of genu recurvatum is voluntarily exaggerated to allow the patient to stand and even to walk.

When the knee is hyperextended (fig. 129) the axis of the thigh runs obliquely inferiorly and posteriorly and the active force (f) can be resolved into a vertical vector (v) representing the body weight acting on the leg, and a horizontal vector (h), which points posteriorly and so tends to accentuate hyperextension. The more oblique posteriorly is the direction of the force (f), the greater is the vector (h) and the more stretched are the posterior ligaments. Therefore if the genu recurvatum is too severe, these ligaments are eventually overstretched and a vicious circle is set up with further accentuation of the genu recurvatum.

Though limitation of hyperextension of the knee is not provided by bony contact, as with the elbow, yet it is no less efficient (fig. 130) and depends essentially on the capsule and the related ligaments and secondarily on the periarticular muscles.

The capsule and its related ligaments consist of:

the posterior capsular ligament (fig. 131);

the collateral ligaments and the posterior cruciate ligament (fig. 132).

The posterior aspect of the capsule (fig. 131) is strengthened by powerful fibrous bands. On either side, in relation to the femoral condyles, the capsule is thickened to form the 'condylar plates' (1), which give attachment to the heads of the gastrocnemius on their deep surfaces. From the styloid process of the fibula radiates a fan-shaped ligament, *the arcuate ligament of the knee*, which contains two bands:

the *lateral band* or the short lateral ligament of Valois, which blends with the lateral 'condylar plate' (2) and its sesamoid bone (3);

the *medial band* radiates medially and its lowermost (4) fibres form the *popliteal tunnel*, which straddles the popliteus tendon as it goes into the capsule (white arrow).

On the medial side the posterior aspect of the capsule is reinforced by the *oblique popliteal ligament* (5), which is formed by recurrent fibres of the semimembranosus tendon (6) and fans out to be inserted into the lateral 'condylar plate'.

All these ligamentous structures of the posterior aspect of the joint are stretched during hyperextension (fig. 132), especially the '*condylar plates*' (1). As shown previously, during extension the *lateral* (7) and *medial* (8) *collateral ligaments* are stretched. *The posterior cruciate ligament* (9) is also stretched during extension. It is easy to establish that during hyperextension the upper attachments (A, B, C) of these structures move anteriorly about the point O as centre.

Finally the **flexor muscles** (fig. 133) play an *active* part in limiting extension: the *three medial tibial muscles* (10), which course behind the medial femoral condyle; the biceps (11) and also the *gastrocnemius* (12). They check extension insofar as they are stretched by dorsiflexion of the ankle.

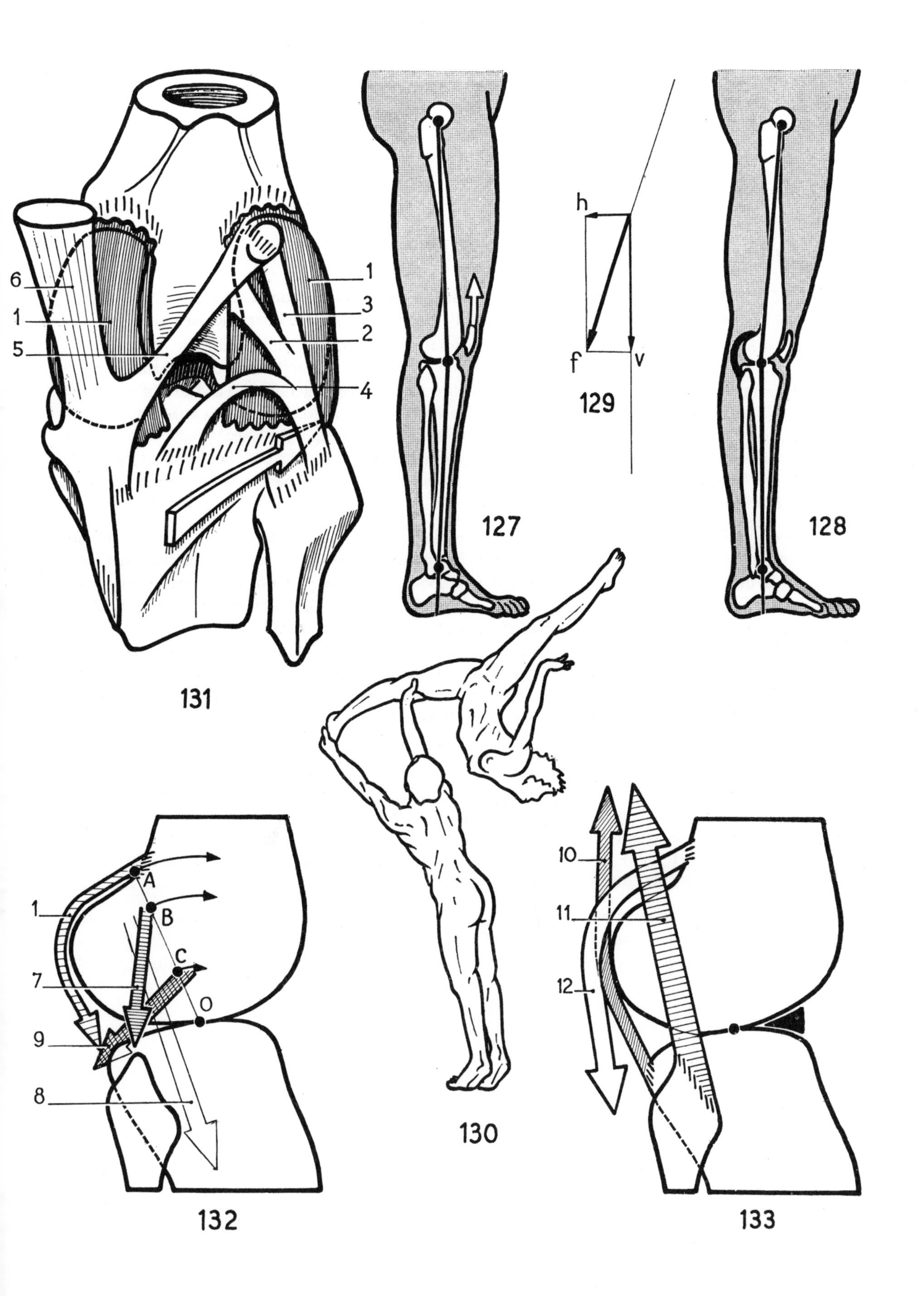

6
1
5
1
3
2
4
h
f
v
129
127
128
131
1
A
B
C
7
O
9
8
10
11
12
130
132
133

THE CRUCIATE LIGAMENTS OF THE KNEE

When the joint is opened anteriorly (fig. 134, according to Rouvière) it becomes obvious that the cruciate ligaments lie in **the centre of the joint** being largely contained within the intercondylar notch.

The first ligament to be seen is the **anterior cruciate ligament** (1), which is attached (fig. 135 (5), according to Rouvière) to the anterior intercondylar fossa of the tibia, along the edge of the medial condyle and between the insertion of the anterior horn of the medial meniscus anteriorly (7) and that of the lateral meniscus (8) posteriorly (see also fig. 64). *It runs obliquely superiorly and laterally* and is attached above (1) (fig. 136, according to Rouvière), to a narrow patch on the internal aspect of the lateral condyle of the femur which extends vertically above and along the edge of the articular cartilage (fig. 64). The ligament has a *more anterior attachment to the tibia* and *a more lateral attachment to the femur* than its fellow; hence the name anterolateral ligament is more apposite.

In the depths of the intercondylar notch behind the anterior cruciate can be seen (fig. 134) **the posterior cruciate ligament** (2). It is attached (fig. 135) to the posterior part of the posterior intercondylar fossa of the tibia, overlapping (according to Rouvière, figs. 136 and 137) the posterior rim of the upper surface of the tibia (fig. 64). Its *tibial insertion* (fig. 135) is therefore placed well posterior to the insertion of the posterior horns of the lateral (9) and medial menisci (10). The ligament *runs obliquely medially, anteriorly and superiorly* (fig. 137: knee flexed 90°) to be inserted (2) into the depths of the intercondylar notch (fig. 138, according to Rouvière) and also to a patch on the edge of the lateral surface of the medial condyle along the line of the articular cartilage (fig. 67). Therefore this ligament has a *more posterior attachment to the tibia* and *a more medial attachment to the femur* than its fellow; hence the more suitable name of the posteromedial ligament.

The posterior cruciate is constantly accompanied by the **meniscofemoral ligament** (3), which is attached below to the posterior horn of the lateral meniscus (figs. 135 and 136): it soon clings (fig. 137) to the anterior surface (usually) of the cruciate (fig. 134) and runs with it to a common insertion into the lateral surface of the medial condyle. Occasionally a *similar ligament* is present in relation to the medial meniscus (fig. 135): a few fibres (12) of the anterior cruciate are inserted into the anterior horn of the medial meniscus near the insertion of the transverse ligament (11).

The cruciate ligaments *touch each other* (fig. 138: the cruciates have been sectioned near their femoral end) on their axial borders, with the anterior running lateral to the posterior ligament. They do not lie free within the joint cavity but are *lined by synovium* and they have important relations with the capsule.

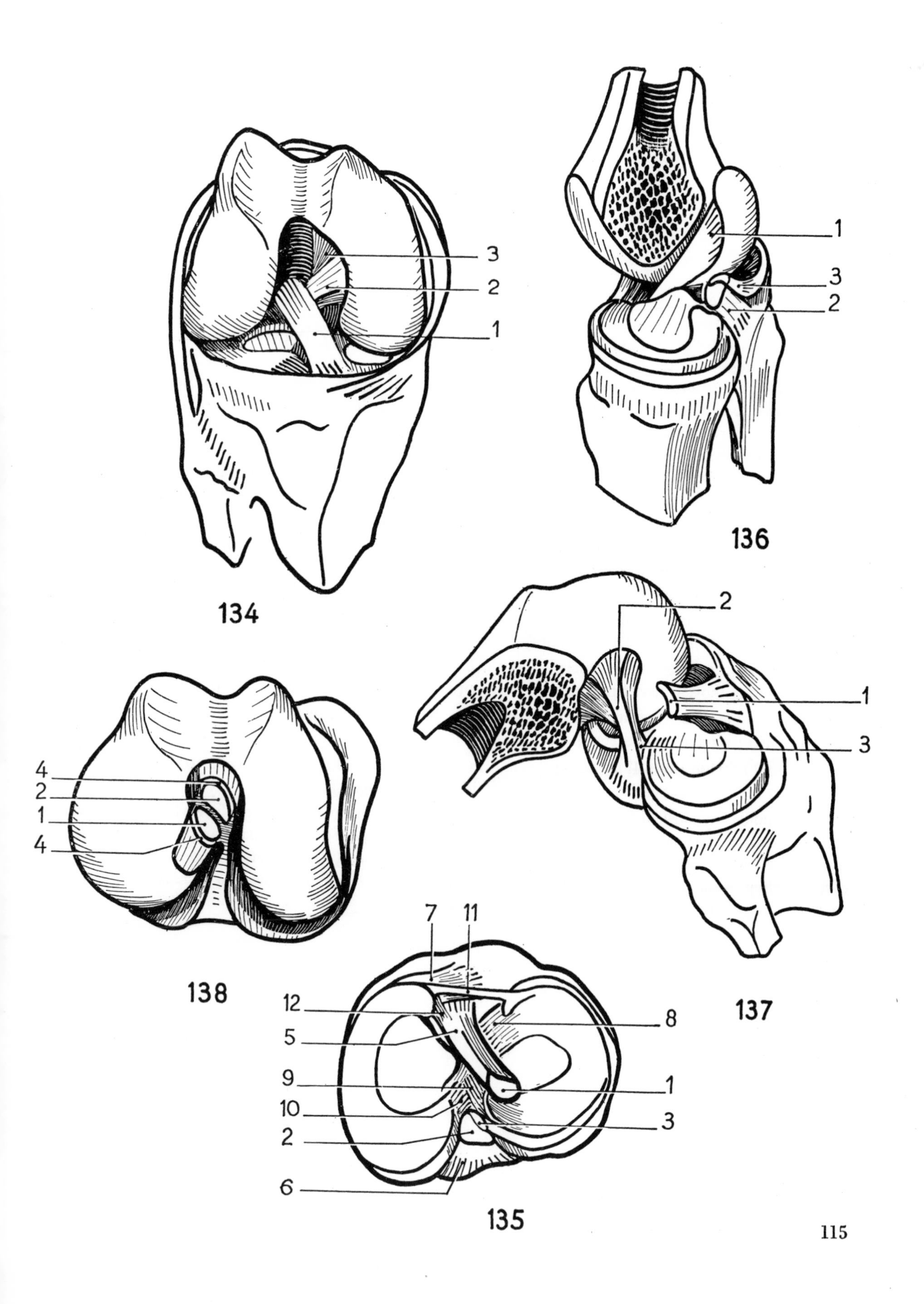
3
2
1
134
1
3
2
136
2
1
3
137
4
2
1
4
138
7
11
12
8
5
9
1
10
3
2
6
135

RELATIONS OF THE CAPSULE AND THE CRUCIATE LIGAMENTS

The cruciates are so intimately related to the capsule that they can be considered as **actual thickenings of the capsule** and, as such, as integral parts of the capsule. It has been shown (p. 92) how the capsule dips into the intercondylar notch to form a double-layered partition along the axis of the joint. It was said earlier, for the sake of convenience, that, to a first approximation, the attachment of the capsule (fig. 139) was such as to make the tibial insertions of the cruciates extracapsular. In fact, the *capsular attachment passes through the attachments* of the cruciates and the thickenings of the capsule, formed by the cruciates, simply stand out on its external surface, i.e. between the two layers of the partition.

Figure 140 (seen from behind and inside: after removal of the medial condyle and section of part of the capsule) shows the **anterior cruciate ligament** clearly applied to the external layer of the capsular partition (the posterior cruciate is not included).

Figure 141 (seen from behind and outside; same as fig. 140) shows the **posterior cruciate** applied to the internal layer of the capsular partition.

Note that all the fibres of the cruciates do not have the same length or the same direction (fig. 143); therefore during movements of the knee they are not all stretched at the same time (p. 120).

These diagrams also illustrate the 'condylar plates', left untouched over the medial condyle (fig. 141) and partially resected over the lateral condyle (fig. 140).

A **coronal section** (fig. 142), passing through the posterior part of the condyles, illustrates the division of the joint cavity (femur and tibia have been artificially pulled apart):

In the middle the capsular partition, thickened by the cruciate ligaments, divides the cavity into a lateral and a medial compartment, this partition is extended anteriorly by the infrapatellar pad (p. 94); each compartment is in turn divided into two storeys by the meniscus, the upper or *suprameniscal* storey corresponding to the line of contact between femur and meniscus, and the lower or *inframeniscal* storey corresponding to the line of contact between tibia and meniscus.

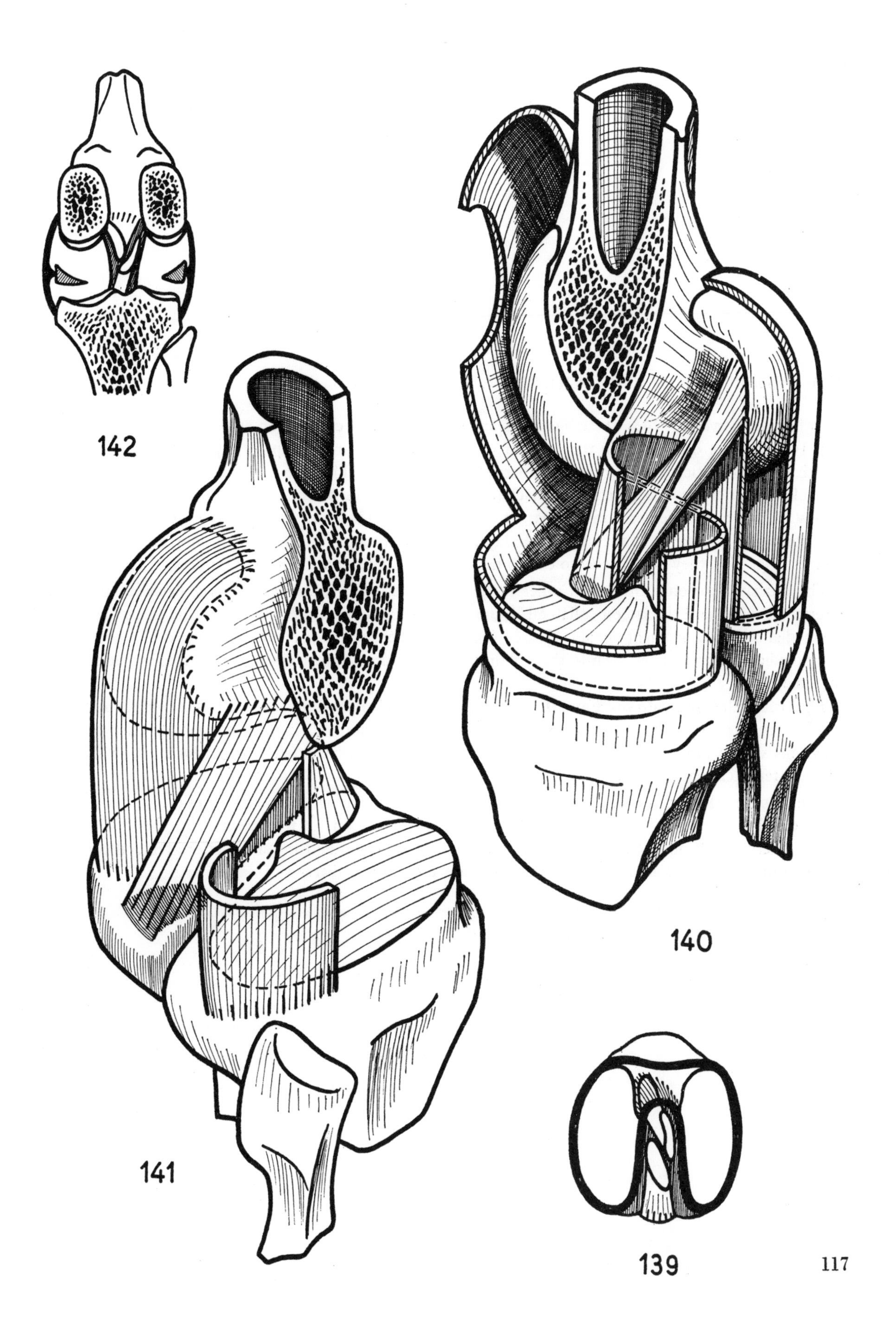
142
141
140
139

THE DIRECTION OF THE CRUCIATE LIGAMENTS

Seen in perspective (fig. 143), these ligaments appear to be in fact **crossed in space**. In the *sagittal* plane (fig. 144) they are crossed, with the anterior cruciate (AC) running obliquely superiorly and *posteriorly* and the posterior cruciate (PC) running superiorly and *anteriorly*. They are also crossed in the *frontal* plane (fig. 145, posterior view) as their tibial attachments (black dots) lie on the anteroposterior axis of the joint (arrow S) while their femoral insertions are 1·7 cm. apart. Thus the posterior cruciate runs obliquely superiorly and *medially* and the anterior obliquely superiorly and *laterally*. In the horizontal plane, by contrast (fig. 159), they run *parallel* to each other and are in contact at their axial borders.

The cruciates not only cross each other in space but also the *ipsilateral collateral ligament*: thus the anterior ligament and the lateral collateral ligament are crossed (fig. 146) and the posterior ligament and the medial collateral ligament are also crossed (fig. 147). Therefore each of these ligaments alternates with its immediate neighbour as regards the obliquity of its course (when taken in order mediolaterally or lateromedially).

The cruciate ligaments **do not have the same angle of inclination** (fig. 144). Thus in full extension the anterior cruciate (AC) is more vertical while the posterior cruciate (PC) is more horizontal. Their femoral insertions show a similar difference: thus the insertion of the posterior cruciate is horizontal (b) while that of the anterior cruciate is vertical (a).

The cruciates **also exhibit a constant length ratio**: in every individual the posterior cruciate is shorter, being equal to three-fifths of the anterior cruciate. This ratio is one of the essential features of the knee and it determines at once the mechanism of action of the cruciates and the shape of the femoral condyles (see p. 120). Also in the adult the tibial insertions are about 5 cm. apart.

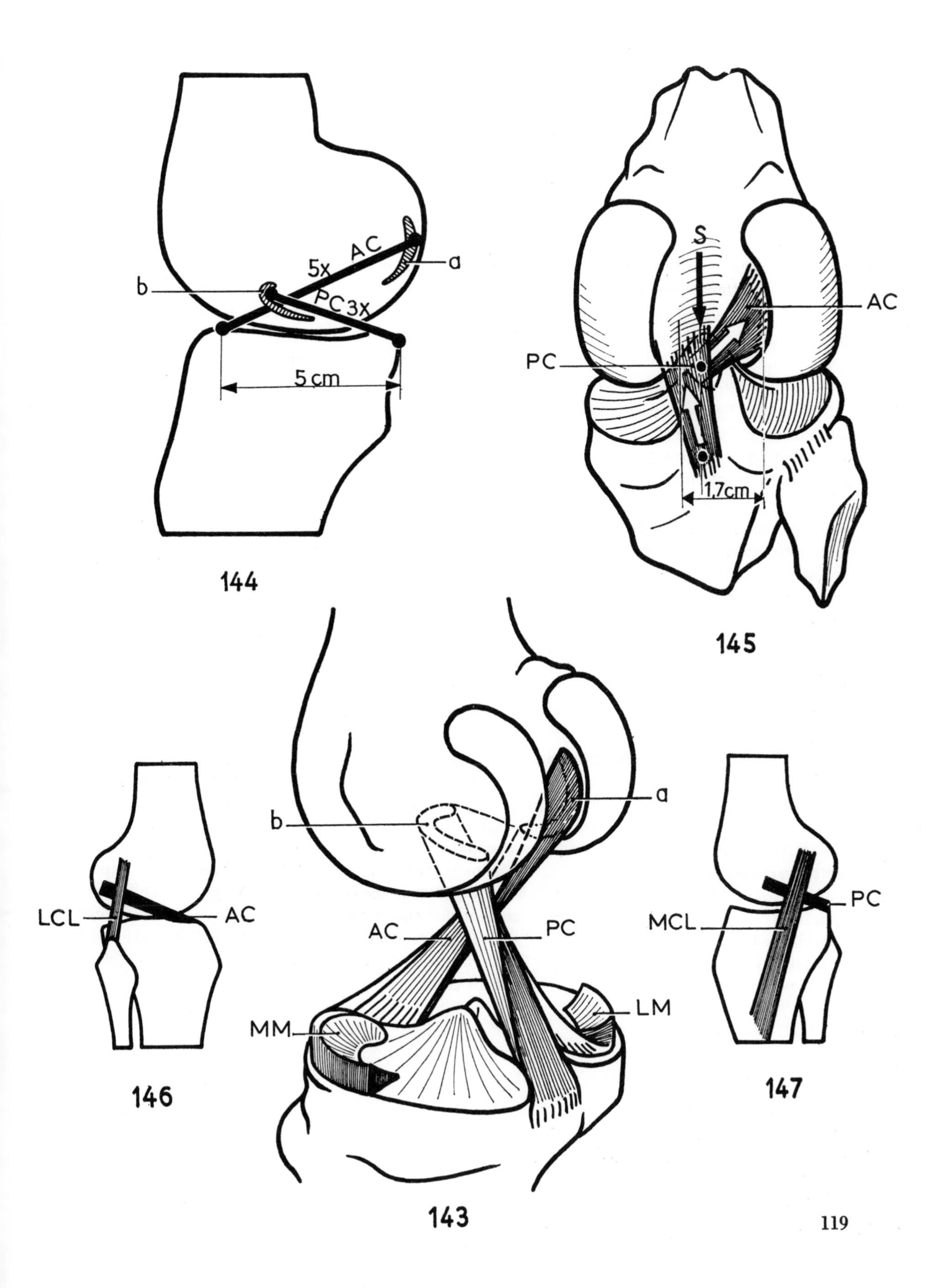
AC
5x
a
b
PC3x
5 cm
144
S
AC
PC
1,7cm
145
a
b
LCL
AC
AC
PC
PC
MCL
MM
LM
146
143
147

THE MECHANICAL ROLE OF THE CRUCIATE LIGAMENTS

These ligaments stabilise the knee in the anteroposterior direction and allow the joint to work as a hinge while keeping the articular surfaces in contact.

The role of the ligaments can be illustrated by a *mechanical model* (fig. 148). Two cardboard sheets (A and B) are attached to each other by ribbons of paper (ab and cd) running from the end of one sheet to the opposite end of the other sheet. These sheets can thus be tilted relative to each other about two hinges, when a and c and b and d coincide but *they cannot slide one on the other.*

The cruciate ligaments have a roughly similar arrangement and function, except that they are not of the same length and the lengths ad and db are not equal. Under these conditions, there are not only two points around which flexion can occur but *a whole series of such points lying on the posterior curve of the condyle.* But, as with the mechanical model (fig. 148), *no anteroposterior sliding is possible* (p. 122).

Starting from the straight position (fig. 150) flexion causes the femoral surface cb to be tilted and the posterior cruciate (fig. 151, cd) is seen to rear itself up, while the point of crossing of the ligaments slides posteriorly and the anterior cruciate (ab) comes to lie horizontal. By lying flat on the tibia the anterior ligament cleaves the intercondylar eminence like a bread knife (fig. 154): during flexion it comes to rest between the two tibial tubercles. When the knee is flexed to 90° (fig. 152), the anterior cruciate (ab) is completely horizontal while the posterior cruciate (cd) becomes vertical. In full flexion (fig. 153); the anterior cruciate (ab) is slackened. During hyperextension (fig. 149) both cruciates are stretched: the anterior cruciate (ab) supports the vault of the intercondylar notch (c); the change in the position of the posterior cruciate is shown on the next page.

The degree of tension developed in the cruciate ligaments during flexion and extension is still a moot point. Roud (1913) maintains that some of their fibres, which are of unequal length, are *always in a stage of tension* (p. 116). On the other hand Strasser (1917) claims, with the help of a mechanical model, that they cannot be stretched simultaneously: the anterior cruciate (AC) is stretched during extension and the posterior cruciate (PC) during flexion. It appears that Roud is right for the following two reasons: firstly, in a normal knee there is *no sliding movement* whatever its position; secondly, with the use of Strasser's model, viewed from a different angle, we have demonstrated that *the profile of the posterior part of the femoral condyle represents exactly the curve joining all the various positions of the tibia plateau* from full flexion to full extension. This proves that *the two ligaments do not change length* while the condyles stay in contact with the tibial plateau. This also illustrates the important concept that **the shape of the condyles is geometrically determined** by the length of the cruciates, their length ratio and the arrangement of their insertions. By modifying the length of the cruciates, their length ratio and their position one can trace a series of different curves which are different but related to that of the condyles.

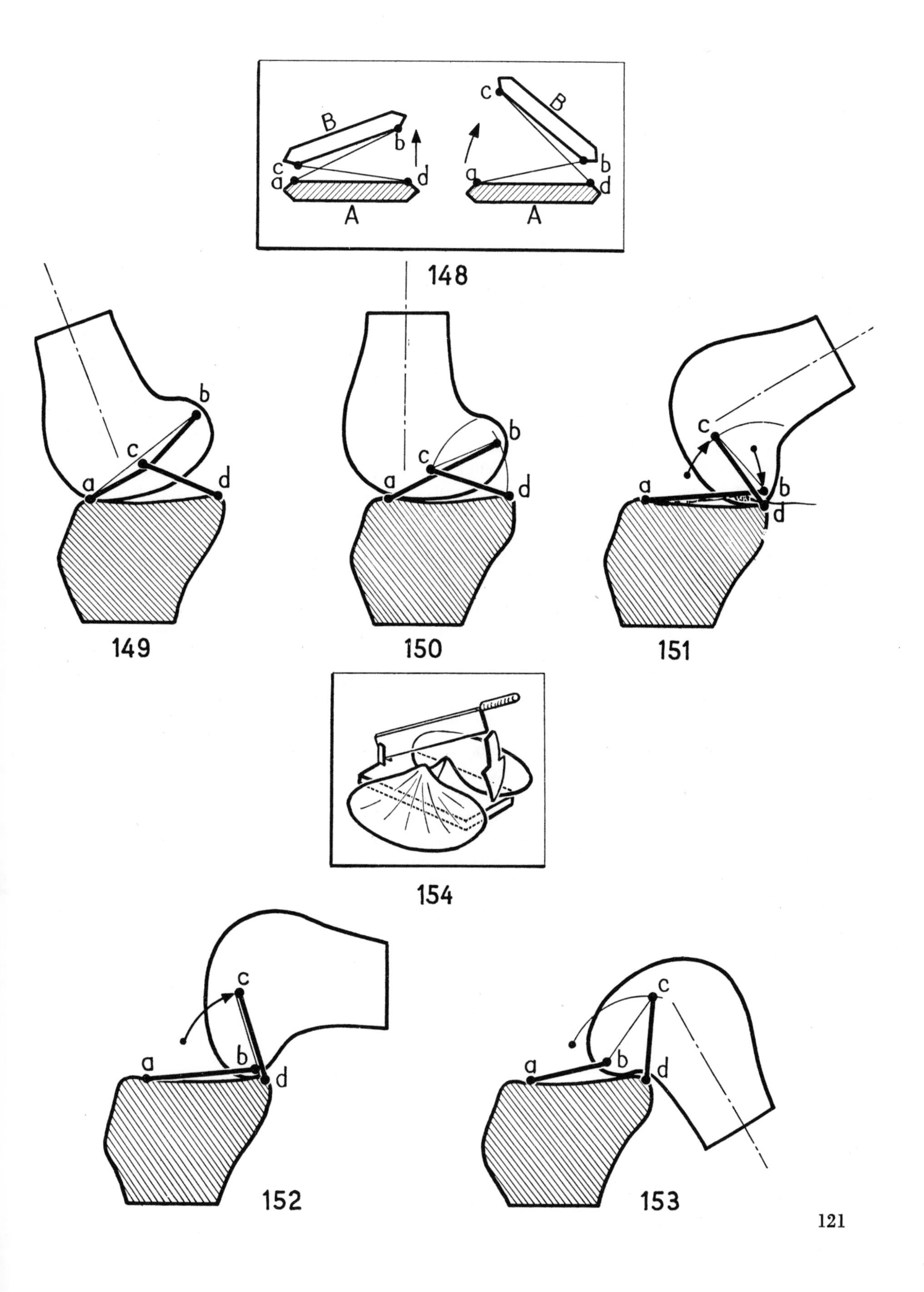
B
c
a
b
d
A
c
B
b
a
d
A
148
b
c
a
d
149
b
c
a
d
150
c
a
b
d
151
154
c
a
b
d
152
c
a
b
d
153

We have seen previously (p. 88) that the movement of the femoral condyles on the tibial condyles combines rolling and sliding. Rolling is easily understood but sliding is not so readily explained in a joint with such a poor degree of interlocking. These are **active mechanisms** involved: the extensors pull the tibia forward underneath the femur during extension (p. 126) and conversely the flexors draw the tibial plateau posteriorly during flexion. But, when these movements are studied on an anatomical preparation, certain **passive mechanisms**, especially the cruciate ligaments, assume paramount importance. *It is the cruciates that pull back the femoral condyles and make them slide on the tibial plateau in a direction opposite to their rolling movement.*

Starting (fig. 155) from the position of extension (I), if the condyle rolled without sliding, it ought to recede to position (II) and the femoral insertion b of the anterior cruciate ab should come to rest in b′, i.e. after covering the distance bb′. However, the point b can only move on a circle with centre a and radius ab (assuming the ligament to be inextensible) so that the real excursion of b is not bb′ but bb″, which corresponds to the position (III) of the condyle lying anterior to position (II) by a distance e. During flexion, the anterior cruciate is called into action and pulls back the condyle anteriorly. It can therefore be said that **during flexion the anterior cruciate is responsible for the sliding movement of the condyle anteriorly**, which occurs while the condyle rolls posteriorly.

The role of the posterior cruciate in extension can be similarly demonstrated (fig. 156). The condyle rolls without sliding from position I to position II and is pulled back posteriorly by the posterior cruciate cd; the distance covered by its femoral attachment c is not cc′ but cc″ lying on a circle with centre d and radius dc. Therefore it follows that the condyle slides posteriorly over a distance f to reach position (III). **During extension the posterior cruciate is responsible for the sliding movement of the condyle posteriorly**, which occurs while the condyle rolls anteriorly.

Abnormal sliding movements are said to occur when the tibia slides anteroposteriorly underneath the femur in the manner of a drawer. These are exhibited with the knee flexed at right angles: if the tibia can be displaced posteriorly on the femur (white arrow) this sliding movement posteriorly is considered abnormal (fig. 157). It is easy to understand that normally such a movement is impossible because of the resistance provided by the posterior cruciate: if it is torn (black arrow) posterior displacement can take place with loss of joint stability in the anteroposterior direction. It is useful to note that a lesion of the POSTERIOR cruciate leads to POSTERIOR displacement.

Similarly (fig. 158) one can show that a lesion of the ANTERIOR cruciate leads to abnormal ANTERIOR displacement.

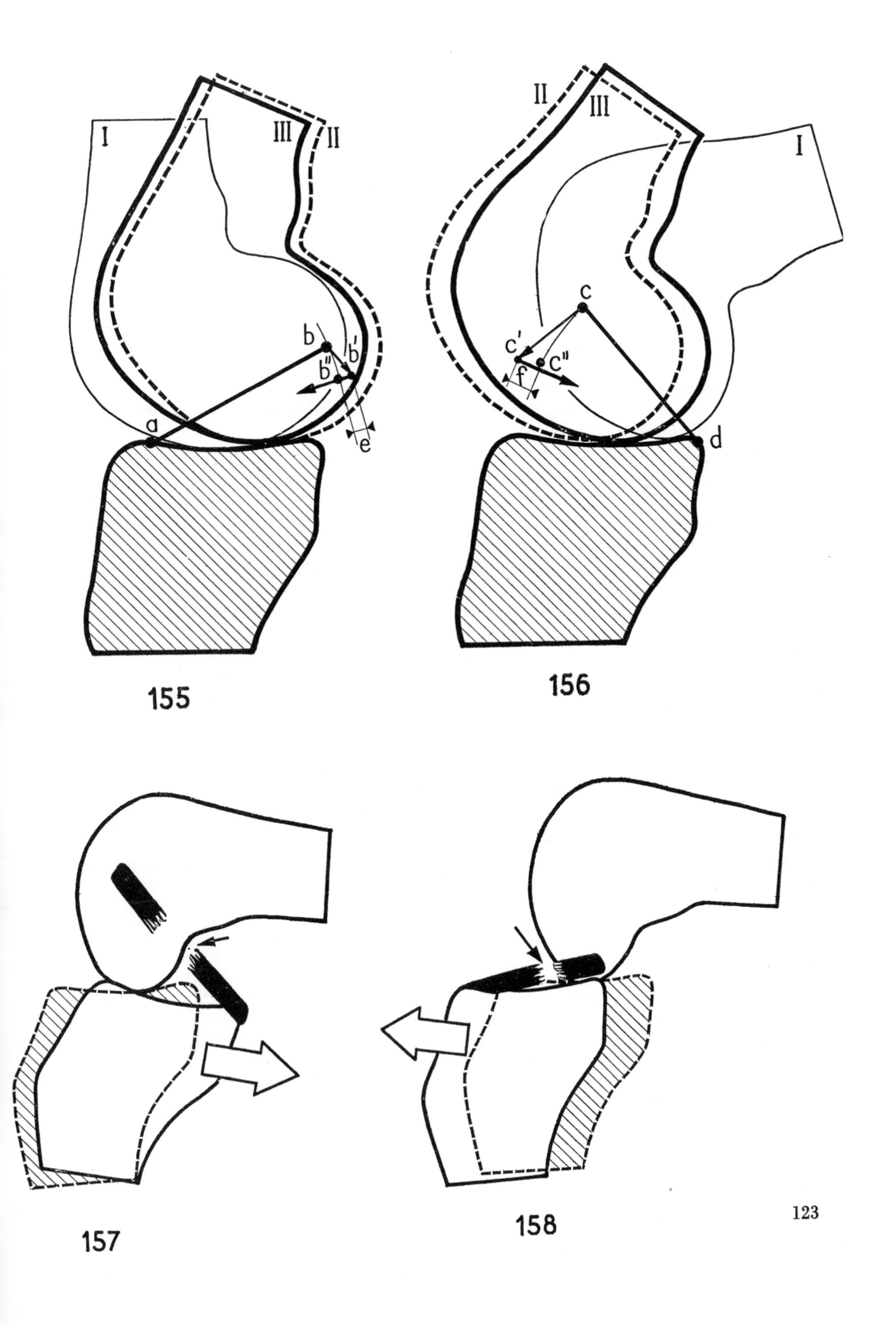
I
III
II
b
b'
b''
a
e
155
II
III
I
c
c'
c''
f
d
156
157
158

THE ROTATIONAL STABILITY OF THE KNEE DURING EXTENSION

We know already that movements of axial rotation can only occur when the knee is flexed. On the other hand, in full extension axial rotation is impossible, **being prevented by the tension of the collateral and cruciate ligaments**:

When seen from above (fig. 159: the condyles are taken to be transparent) the cruciate ligaments run obliquely: the anterior ligament (white) inferiorly, anteriorly and medially; the posterior ligament (black) inferiorly, posteriorly and laterally. They are *'coiled' anticlockwise* round the axis of rotation (right knee). If the tibia rotates laterally under the femur (fig. 160) i.e. clockwise, the cruciates become separated and vertical (fig. 161) and the tibia moves away slightly from the femur. Therefore *lateral rotation relaxes the cruciates*. Conversely, if the tibia rotates medially (fig. 162) i.e. anticlockwise, the cruciates come into contact along their axial borders and become coiled round each other (fig. 163), so that they are effectively shortened and the tibia is pressed against the femur. Therefore *medial rotation tightens the cruciates*. As the femoral condyles are already in contact with the tibial condyles (figs. 161 and 163: the femur and tibia have been deliberately separated) it is easy to understand why **the cruciates prevent medial rotation when the knee is extended.**

A similar line of reasoning can be developed to explain the role of the collateral ligaments. These run obliquely (fig. 164) and they are *'coiled' clockwise* (right knee) around the axis of rotation. If the tibia is medially rotated (fig. 165), i.e. anticlockwise, the collateral ligaments become more vertical (fig. 166) and the tibia moves away from the femur: *medial rotation relaxes the collateral ligaments*. If the tibia is laterally rotated (fig. 167), these become more oblique (fig. 168) and the tibia is pressed more strongly against the femur: *lateral rotation tightens the collateral ligaments*. Therefore it follows that **the collateral ligaments prevent lateral rotation of the knee in extension.**

The rotational stability of the knee is thus secured by the collateral and cruciate ligaments.

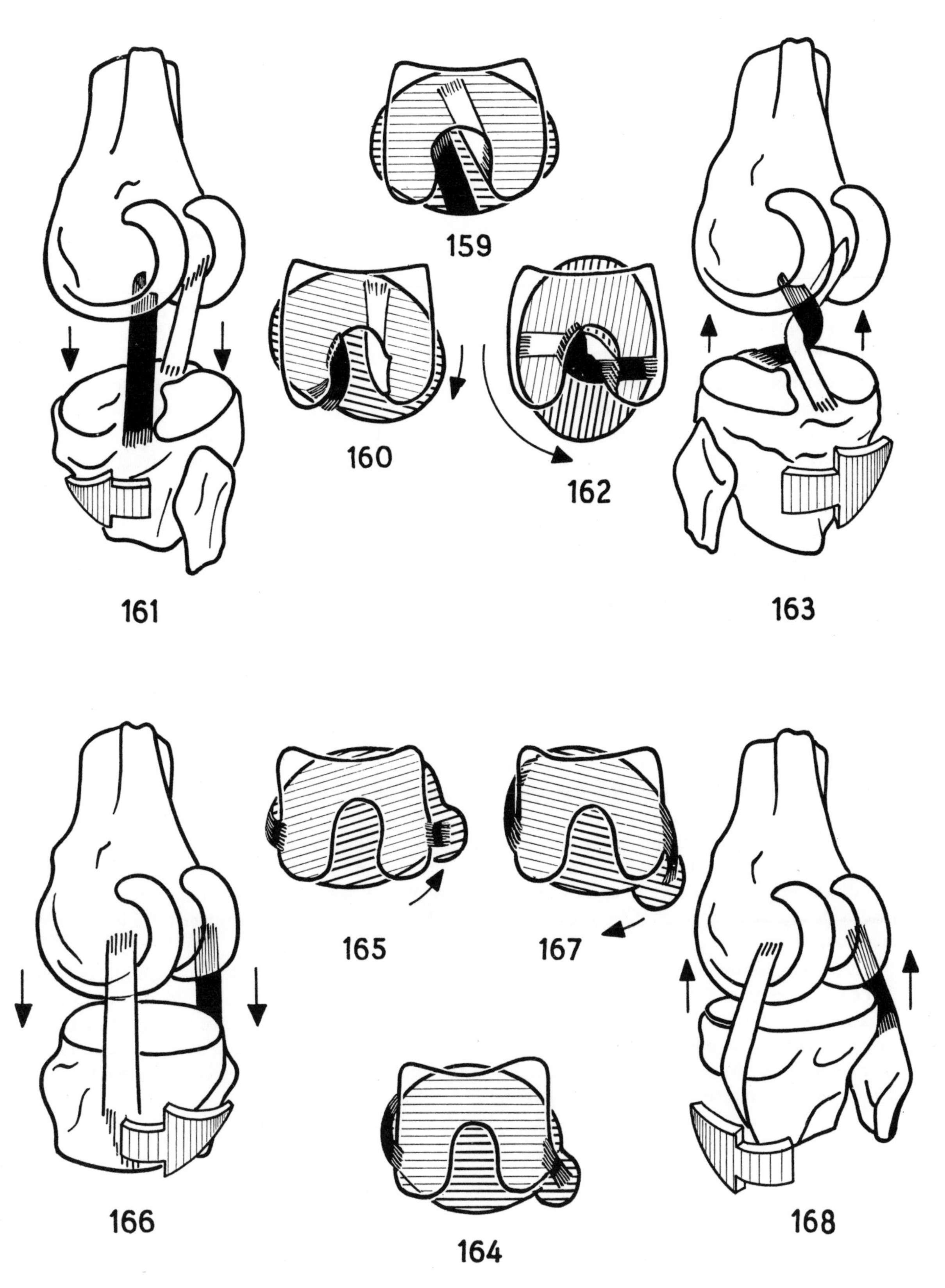
159
160
162
161
163
165
167
166
168
164

THE EXTENSOR MUSCLES OF THE KNEE

The **quadriceps femoris** is **the extensor muscle of the knee**. It is a *powerful muscle*: its active cross-sectional area is 148 cm^2 and, as *it shortens by a distance of* 8 *cm.*, it develops a force equivalent to 42 kg. weight. It is *three times stronger* than the flexors, as can be expected from the fact that it counteracts the effect of gravity. We have already seen, however, that when the knee is hyperextended the quadriceps is not required for maintenance of the erect position (p. 112) but, as soon as flexion is initiated, the quadriceps is strongly thrown into action so as to prevent a fall resulting from knee flexion.

The quadriceps (fig. 169), as suggested by its name, consist of four muscles which are inserted by a common tendon into the anterior tuberosity of the tibia. Three of its muscles are monoarticular—the *vastus intermedius* (1), *vastus lateralis* (2) and *vastus medialis* (3)—and the fourth, the *rectus femoris* (4), is biarticular (for its function, see p. 128). These three monoarticular muscles are exclusively extensors of the knee, but the vastus lateralis and vastus medialis also possess a lateral component of force. Therefore a balanced contraction of these vasti produces a force acting along the axis of the thigh but, if there is an imbalance of these muscles, the patella can be abnormally pulled to one side; this is one of the many mechanisms implicated in the recurrent dislocation of the patella (p. 102).

The patella is **a sesamoid bone** embedded in the extensor tendon of the knee; its function is to increase the efficiency of the quadriceps *by shifting anteriorly the line of action of its muscular pull*. This is readily demonstrated by studying **the diagram of forces with and without the patella**.

The force Q of the quadriceps, acting on the patella (fig. 170), can be resolved into two vectors: a force Q1, acting towards the axis of flexion and extension and tending to keep the patella pressed against the femur, and a force Q2 acting along the line of the ligamentum patellae. This force Q2, acting on the tibial tuberosity, can also be resolved into *two vectors perpendicular to each other*: a force Q3, acting towards the axis of flexion and extension and keeping the tibia and femur together, and a tangential force Q4 which is *the component effective in extension*, i.e. it moves the tibia anteriorly underneath the femur.

Let us assume that the patella has been removed—i.e. after a patellectomy—and let us proceed as before (fig. 171). The force Q (assuming this one to be equal to the other Q) acts tangentially to the patellar surface of the femur and directly on the tibial tuberosity. It can therefore be resolved into two vectors: Q5, which keeps the tibia pressed against the femur, and Q6, the component effective for extension. Note that the tangential component of Q6 is clearly reduced whereas the centripetal component Q5 is relatively enhanced.

If we now compare the effective forces in these two situations (fig. 172) it is clear that Q4 is about one and a half times greater than Q6: *thus the patella, by raising the quadriceps tendon as on a trestle, indisputably increases* its *efficiency*. It is also evident that, in the absence of the patella, the force of coaptation Q5 is increased but this favourable effect is cancelled by the reduction in the range of flexion secondary to shortening of the extensor tendon. **The patella is therefore a very useful mechanical device** and should only be removed when absolutely necessary.

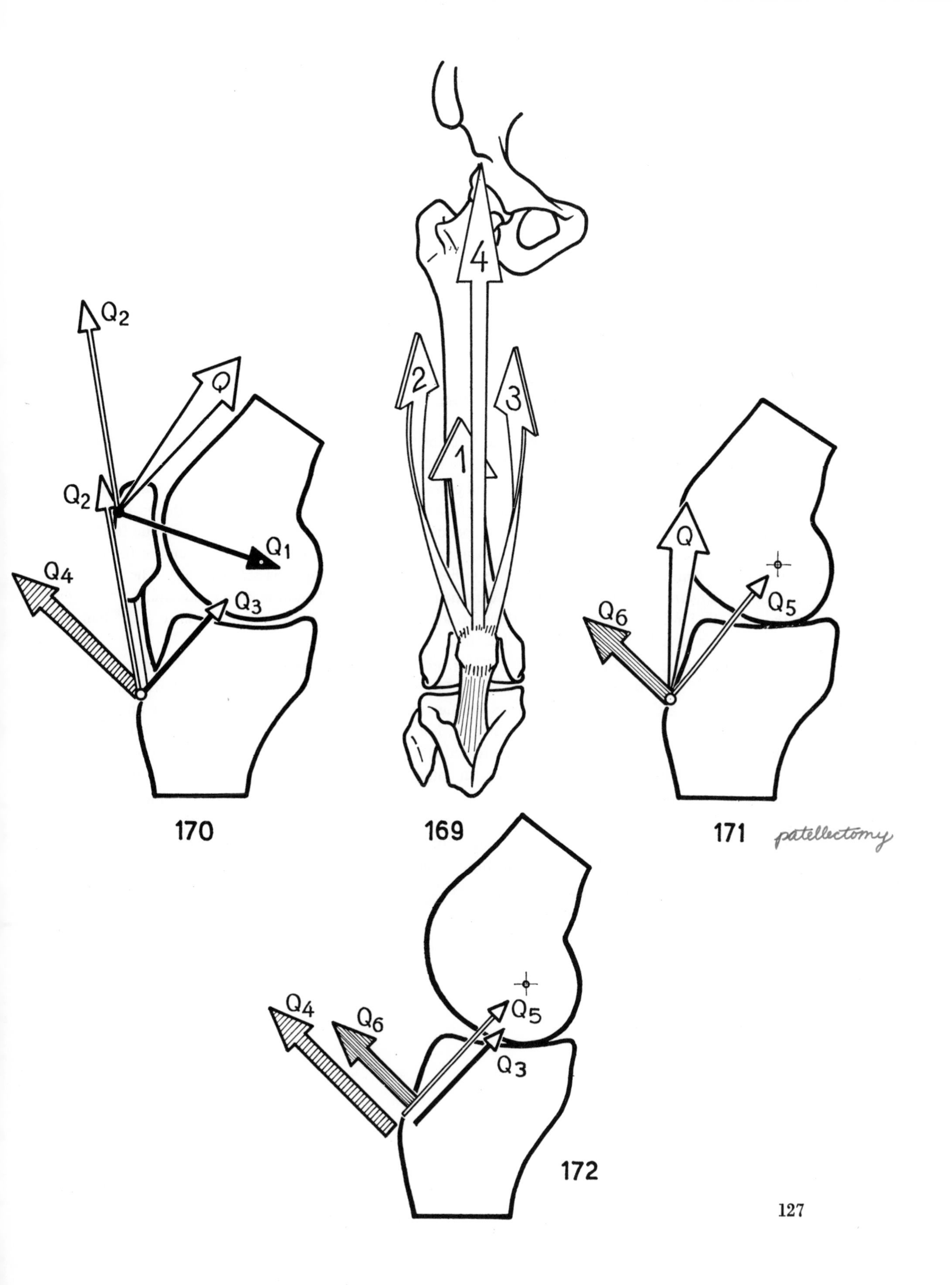

170

169

171

172

THE PHYSIOLOGICAL ACTIONS OF THE RECTUS FEMORIS

The rectus provides only *one-fifth* of the total force of the quadriceps and it cannot by itself produce full extension. But its biarticular nature gives it special significance.

As it runs anteriorly to the axis of flexion and extension of the hip and knee, it is at once a *flexor of the hip* and an *extensor of the knee* (fig. 174), but its efficiency as a knee extensor depends on the position of the hip and conversely its action as a hip flexor depends on the position of the knee. This is due to the fact (fig. 173) that the distance between the anterosuperior iliac spine (a) and the superior end of the patellar surface of the femur is shorter when the hip is flexed (ac) than when it is straight (ab). The difference (e) produces a *relative lengthening* of the muscle when the hip is flexed and the knee bends under the weight of the leg (II). Under these conditions, the vasti are more efficient in extending the knee (III) than the rectus, which is already slackened by hip flexion.

On the other hand, if the hip is extended (IV) from the reference position (I), the distance between the origin and insertion of the rectus (ad) increases by f and this stretches the rectus (i.e. relative shortening) and enhances its efficiency. This also occurs during running or walking when the posterior limb is lifted off the ground (fig. 177): the glutei extend the hip while the knee and ankle are flexed. The quadriceps then works at its best advantage because of the increased efficiency of the rectus. *The gluteus maximus is therefore an antagonist-synergist of the rectus femoris*, i.e. an antagonist at the hip and synergist at the knee.

When one limb moves forward off the ground (fig. 176), the pelvis being supported temporarily on the other hip, the rectus contracts to produce at once flexion of the hip and extension of the knee. Therefore the rectus, as a biarticular muscle, is useful in both phases of walking, i.e. when the posterior limb provides the propulsive thrust and when the anterior limb is moved forwards (Vol. IV).

Finally, knee flexion favours its flexor action on the hip: during a jump with flexed knees (fig. 175) the two recti play an important part in flexing the hip.

a
c
III
f
c
e
b
b
d
IV
II
I
174
173
177
176
175

THE FLEXOR MUSCLES OF THE KNEE

These are lodged in the **posterior compartment of the thigh** (fig. 178); they are the *hamstring muscles* —biceps femoris (1), semitendinosus (2), semimembranosus (3); the *three muscles inserted the medial aspect of the tibia*—the gracilis (4), sartorius (5) and semitendinosus (also part of the hamstrings); and the *popliteus* (p. 132). The gastrocnemius (6 and 7) is practically useless as a knee flexor but is a powerful extensor of the ankle (p. 186).

All these muscles are biarticular with two exceptions: the short head of the biceps and the popliteus which are monoarticular (p. 132). These biarticular flexors therefore also extend the hip simultaneously and their action on the knee depends on the position of the hip.

The sartorius (5) is a flexor, abductor and lateral rotator of the hip and at the same time a flexor of the knee.

The **gracilis** (4) is primarily an adductor and an accessory flexor of the hip; it also flexes the knee and participates in medial rotation of the knee (p. 132).

The **hamstrings** are at once extensors of the hip (p. 50) and flexors of the knee and their action on the knee depends on the position of the hip (fig. 179). When the hip is flexed, the distance ab between the origins and insertions of these muscles increases progressively since the centre of the hip O, around which the femur turns, does not coincide with the point a around which the hamstrings turn. Thus the more the hip is flexed the greater the degree of relative shortening of these muscles and the more *stretched* they become. When the hip is flexed at 40° (II), the relative shortening of the muscles can be partly made up for by *passive* flexion of the knee (ab=ab′). However when hip flexion reaches 90° (III), the relative shortening cannot be wholly compensated even by a 90° flexion of the knee (f= 'residual' shortening). As hip flexion exceeds 90° (IV), it becomes very difficult to keep the knee in full extension (fig. 180): the amount of relative shortening (g) is just about absorbed by the elasticity of the hamstrings, which decreases remarkably with lack of exercise. **When the hamstrings are stretched by hip flexion their efficiency as knee flexors increases**: thus, during climbing (fig. 181) when one lower limb moves forward, flexion of the hip increases the efficiency of the knee flexors. Conversely, knee extension promotes the flexor action of these muscles on the hip: this occurs as one tries to straighten the trunk when it is bent forward (fig. 180) and when, during climbing, the posterior limb moves in front of the other limb.

If the hip is maximally extended (fig. 179, portion V) the hamstrings show a *relative lengthening* (3) and so they lose some of their efficiency (fig. 10) as knee flexors.

These observations stress the *usefulness of the monoarticular muscles* (popliteus and short head of biceps), which have the same efficiency whatever the position of the hip.

The total force produced by the flexors is equivalent to 15 kg. weight, i.e. about one-third of that produced by the extensors.

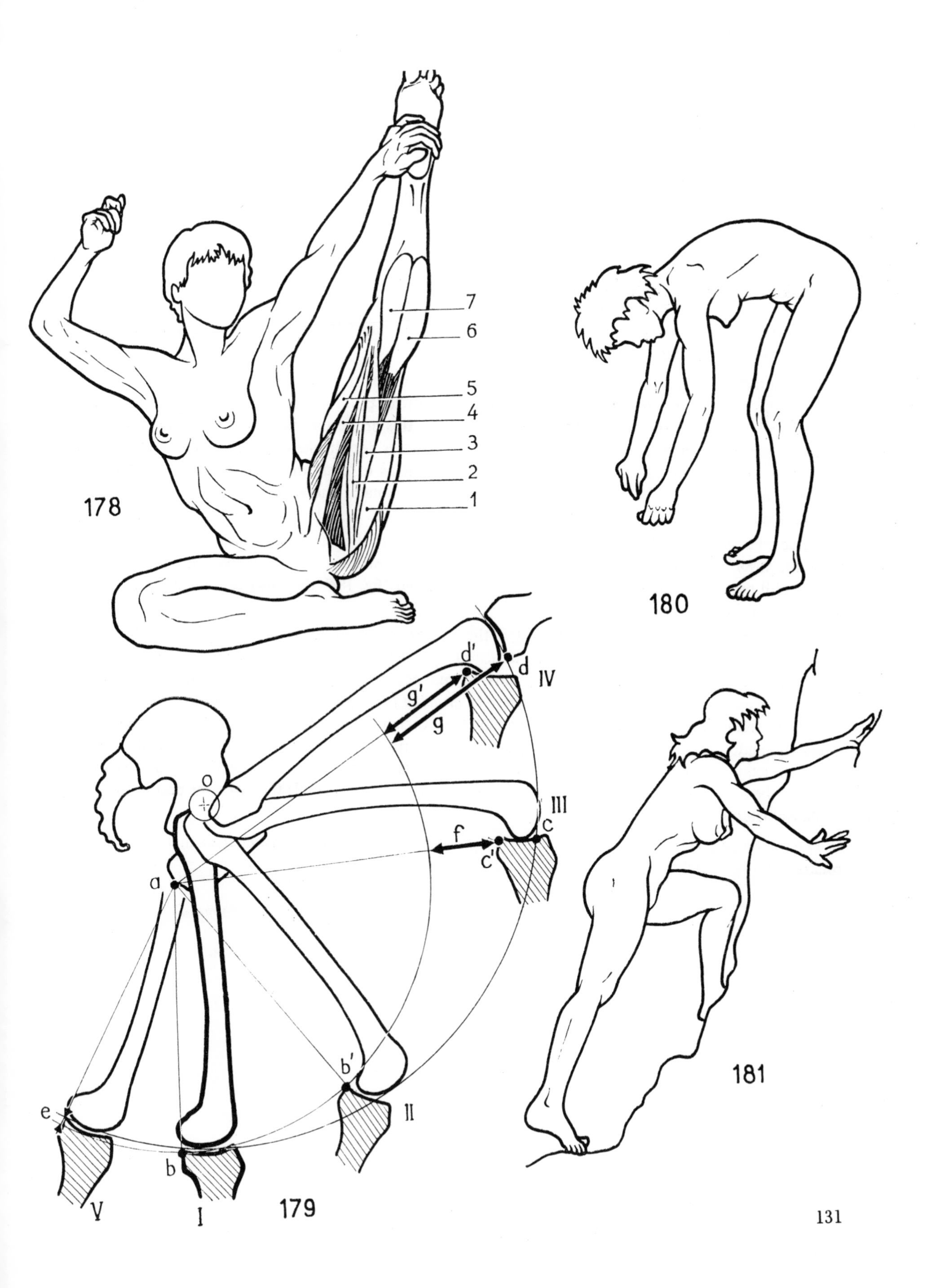
7
6
5
4
3
2
1
178
180
d'
d
IV
g'
g
o
III
f
c
c'
a
181
b'
II
e
b
V
I
179

THE ROTATOR MUSCLES OF THE KNEE

The flexors are also at the same time **rotators** of the knee and they fall into two groups depending on their locus of insertion into the leg bones (fig. 182):

Those attached *lateral* (A) to the vertical axis XX′ of rotation of the knee are **lateral rotators** (fig. 185), i.e. *biceps* (1) and *tensor fasciae latae* (2). When these muscles pull the lateral aspect of the tibial plateau posteriorly (fig. 183) they cause it to rotate so that the tips of the toes face more *laterally*. The tensor fasciae latae is only a flexor and lateral rotator when the knee is flexed; when it is fully extended the muscle loses its rotator action and becomes an extensor, i.e. it helps to 'lock' the knee in extension. The *short head of biceps* (fig. 186, I′) is the only *monoarticular* lateral rotator and so the position of the hip has no effect on its function.

Those attached *medial* (B) to the vertical axis XX′ of rotation of the knee are the *medial rotators* (fig. 185), i.e. the *sartorius* (3), *semitendinosus* (4), *semimembranosus* (5), *gracilis* (6) and *popliteus* (figs. 186, 7). When they pull posteriorly the medial aspect of the tibial plateau (fig. 184) they also cause it to rotate so that the tips of the toes now look *medially*.

The **popliteus** (fig. 188, seen from behind) is the only exception to this general rule. It arises by tendon from the popliteal groove on the lateral surface of the lateral femoral condyle and soon penetrates the *articular capsule*—still outside the synovium—to run between the lateral collateral ligament and the lateral meniscus. It sends a *fibrous expansion to be attached to the posterior edge of the lateral meniscus* and then emerges from the capsule under cover of the cruciate ligament of the knee before reaching its insertion into the posterior aspect of the upper end of the tibia. It is the only *monoarticular* medial rotator and its action is uninfluenced by the position of the hip. Its action is easily illustrated in figure 187 which shows the tibial plateau seen from above: the popliteus (black arrow) pulls the posterior part of the tibial plateau laterally.

The combined power of the medial rotators (2 kg. weight) is only a little greater than that of the lateral rotators (1·8 kg. weight).

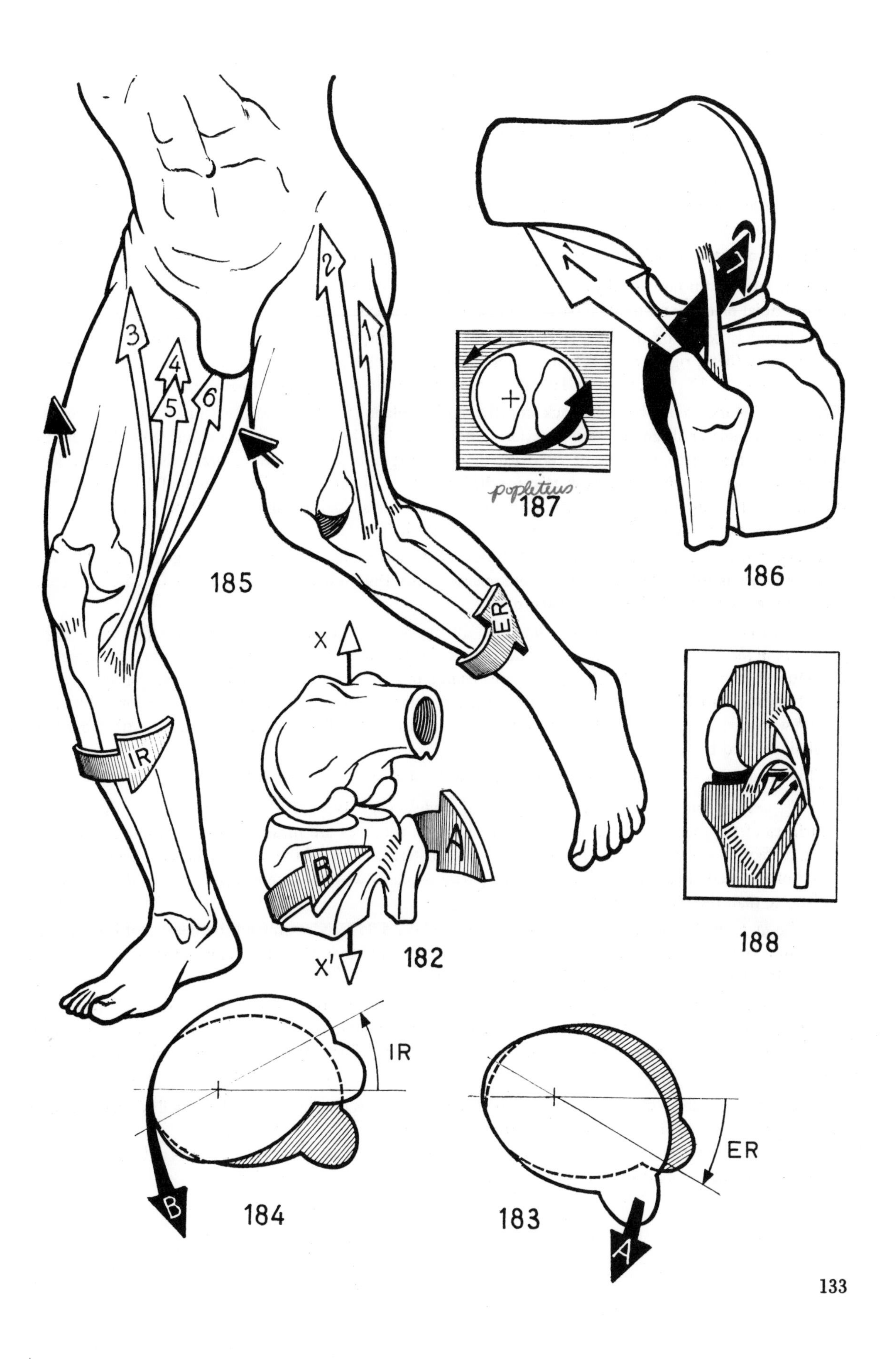

185

186

187

182

188

184

183

THE AUTOMATIC ROTATION OF THE KNEE

It has already been noted (p. 78) that *the terminal phase of extension is associated with a small measure of lateral rotation* and that *the beginning of flexion is always combined with some medial rotation*. These movements of rotation occur *automatically*, i.e. *in the absence of any voluntary movement*.

This **automatic rotation** can be displayed on an anatomical preparation based on Roud's experiment.

Two pins are driven horizontally and transversely (fig. 189, seen from above) through the femoral condyles and the tibial plateau and they lie parallel when the knee is extended.

If the femur is flexed on the tibia (fig. 190), which stays put, the femoral axis runs obliquely posteriorly and medially (here, right knee). When the angle of flexion is 90°, the two pins now form in the horizontal plane an angle of 30°, which 'looks' laterally and posteriorly (45° according to Roud).

When the femoral axis is brought back into the sagittal plane (fig. 191), the tibial pin is now seen to run obliquely mediolaterally and postero-anteriorly, indicating that the tibia has been *medially rotated* under the femur. This pin forms an angle of 20° with the line perpendicular to the femoral axis. Therefore knee flexion is associated with **an automatic medial rotation of 20**°. This discrepancy of 10° results from the fact that the femoral pin (not shown), because of the physiological valgus, is not perpendicular to the axis of the femoral shaft but forms an angle of 80° with it (fig. 3).

This experiment can be reversed: in the position of flexion at 90° the pins are divergent (fig. 190), and, when the knee is fully extended, they become parallel (fig. 189). This shows that knee extension is accompanied by **an automatic lateral rotation.**

Medial rotation of the tibia occurs (fig. 192) during flexion because the *lateral femoral condyle recedes farther than the medial condyle.* In the extended knee the points of contact a and b lie on the transverse axis O*x*. During flexion, the medial condyle recedes from a to a′ (5 to 6 mm.) while the lateral condyle recedes from b to b′ (10 to 12 mm.) and the points of contact during flexion lie on a line O*y* which forms an angle of 20° with O*x*. Therefore, if O*y* is once more to coincide with the transverse plane, the tibia must be medially rotated by 20°.

This unequal posterior movement of the condyles is due to *three mechanisms*:

1. *The unequal length of the profiles of the femoral condyles* (figs. 193 and 194): if the lengths of the articular surfaces of the medial (fig. 193) and lateral (fig. 194) condyles are measured out by rolling them on a flat surface, it is clear that the length bd′ of the posterior curve of the lateral condyle slightly exceeds that of the medial condyle (ac′=bc′). This partly explains why the lateral condyle rolls over a greater distance than the medial condyle.
2. *The shape of the tibial condyles*: the medial femoral condyle recedes only a little because it is contained within a concave tibial surface (fig. 195), while the lateral femoral condyle slides more freely over the posterior slope of the convex tibial surface (fig. 196).
3. *The direction of the collateral ligaments*: when the femoral condyles recede, the medial collateral ligaments is stretched more rapidly (fig. 195) than the lateral collateral ligament (fig. 196); the latter therefore allows the condyle to recede farther because of its obliquity. There are also *two couples of force producing rotation*:

the predominant action of the flexor and medial rotator muscles (fig. 197): the gracilis, sartorius and semitendinosus (black arrow); and the popliteus (white arrow);

the tension of the anterior cruciate at the end of extension (fig. 198): as the ligament comes to lie to the lateral side of the axis of the joint the tension developed in it produces lateral rotation.

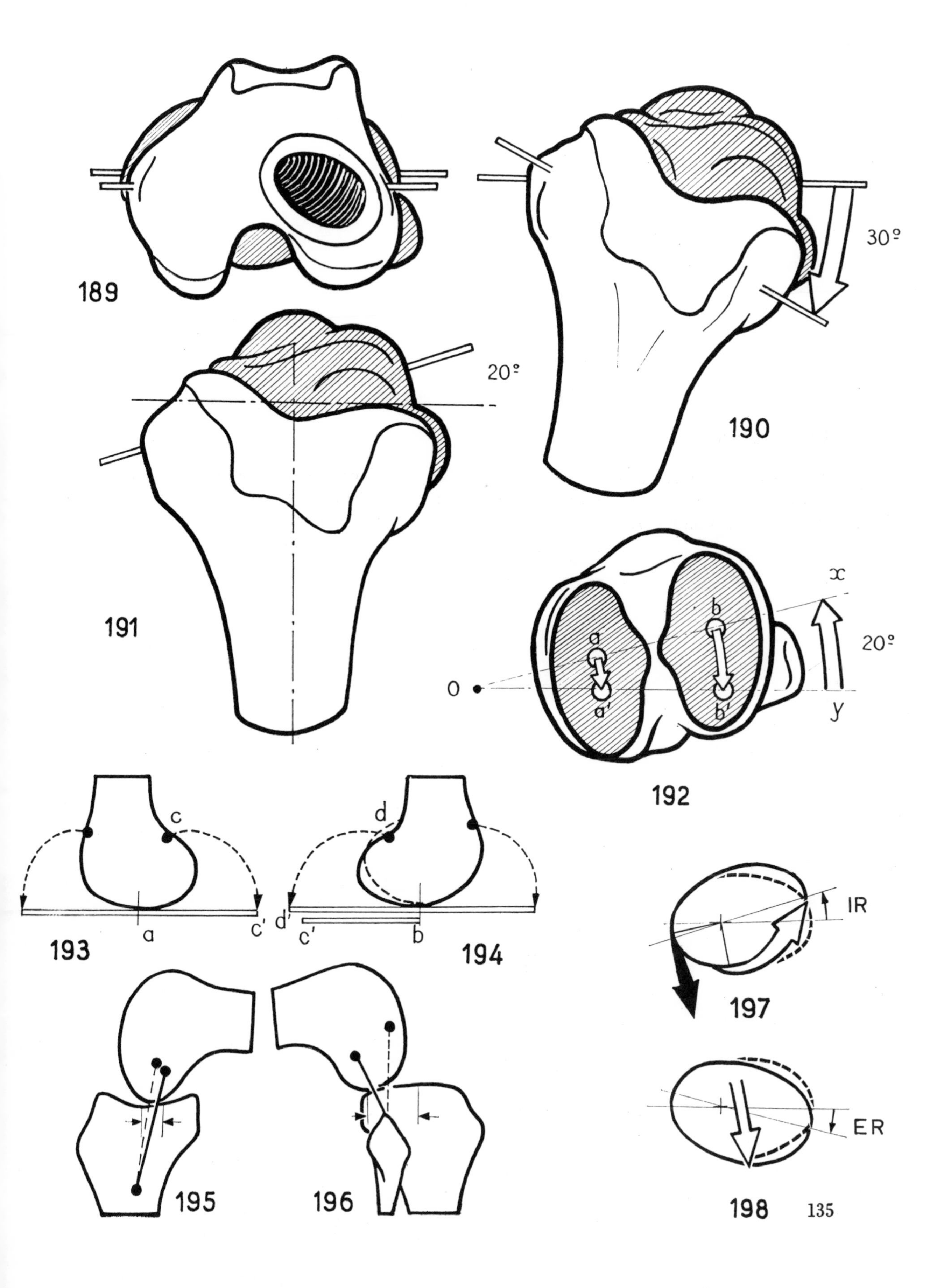
189
190
30º
20º
191
x
b
a
20º
O
a'
b'
y
192
c
d
a
c'
d'
c'
b
193
194
IR
197
ER
195
196
198

THE ANKLE

The ankle or the tibiotarsal joint is the distal joint of the lower limb. It is a **hinge joint** and has therefore only *one degree of freedom.* It controls the movements of the leg relative to the foot, which occur in a *sagittal plane.* These movements are essential for walking on flat or rough ground.

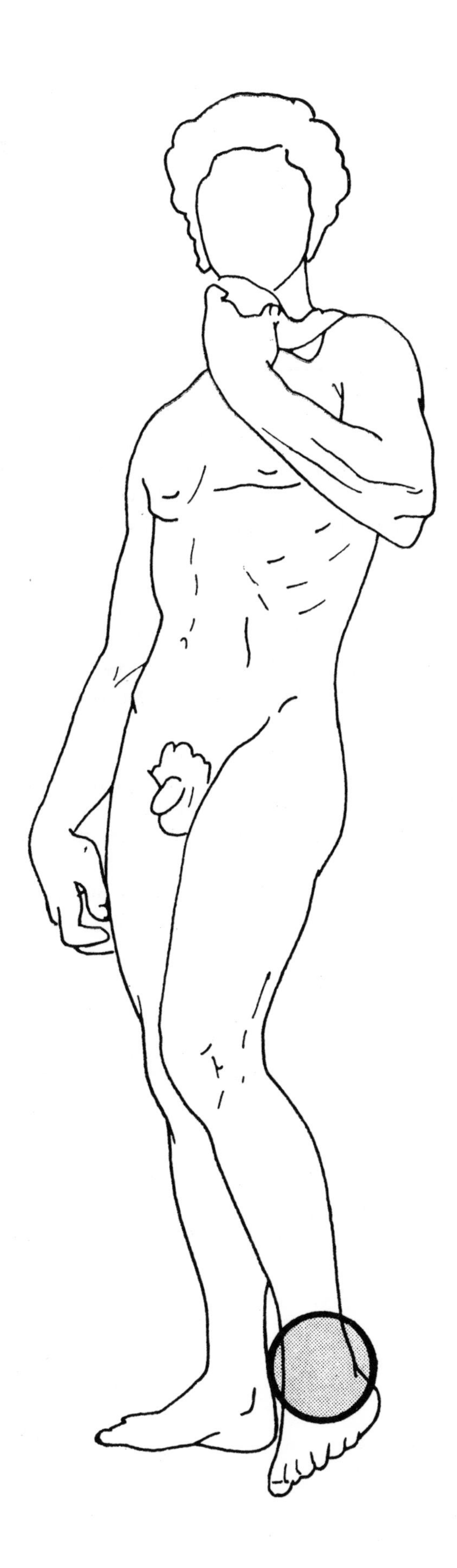

THE JOINT COMPLEX OF THE FOOT

In fact, the ankle is only the most important of the joints of the posterior half of the foot. This series of joints, assisted by axial rotation of the knee, is in practice equivalent to a single joint with three degrees of freedom; it allows the foot to take up any position in space and to adapt to any irregularities of the ground. A certain similarity to the upper limb is evident: the joints of the wrist, assisted by pronation and supination, allow the hand to assume any position in space but the mobility of the hand is much greater than that of the wrist.

The **three main axes** of this joint complex (fig. 1) intersect roughly in the posterior half of the foot. When the foot lies in the position of reference, the three axes are perpendicular one to another. In the diagram, extension of the ankle changes the direction of the Z axis.

The **transverse axis** XX′ passes through the two malleoli and corresponds to the axis of the ankle proper. It lies almost wholly in the frontal plane and controls *the movements of flexion and extension* of the foot (p. 140), which occur in a sagittal plane.

The **long axis of the leg Y** is vertical and controls the *movements of adduction and abduction* of the foot, which take place in a transverse plane. As shown previously (p. 78), these are only possible because of axial rotation of the flexed knee. A smaller proportion of these movements of adduction and abduction depends on the joints of the posterior part of the foot but then they are always associated with movements around the third axis (p. 156).

The **long axis of the foot Z** is horizontal and lies in a sagittal plane. It controls the movements of the sole of the foot and allows it to face inferiorly, laterally or medially. By analogy with the upper limb these movements can be called *pronation* and *supination* respectively.

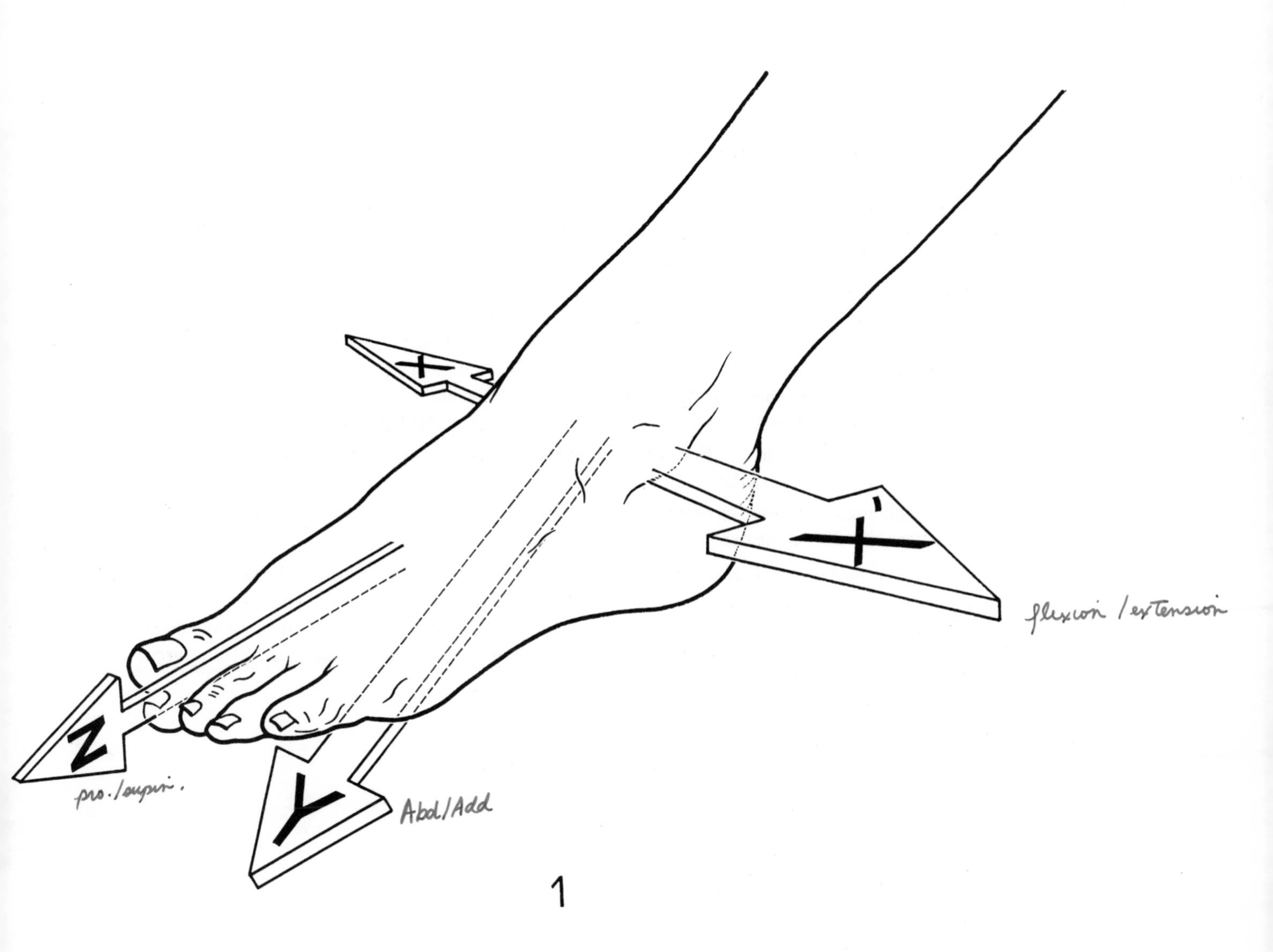
X
X'
flexion / extension
Z
pro. / supin.
Y
Abd/Add
1

FLEXION AND EXTENSION

The **position of reference** (fig. 2) is achieved when the sole of the foot is perpendicular to the axis of the leg (A). From this position, **flexion of the ankle** (B) is the movement which approximates the dorsum of the foot and the anterior surface of the leg; it is also called dorsiflexion.

Conversely, **extension of the ankle** (C) is the movement of the dorsum of the foot away from the anterior surface of the leg so that the foot tends to fall into line with the leg. It is also called plantar flexion but this term is incorrect because flexion always corresponds to the movement of approximation of the segments of a limb and the trunk. In the diagram it is clear that the range of extension is distinctly greater than that of flexion. In measuring these angles, the centre of the ankle is not used as the reference point, as it is simpler to assess the angle between the sole of the foot and the axis of the leg (fig. 3).

When this angle is acute (b), **flexion** is present. Its range is from 20° to 30°. The striped zone indicates the range of individual variations, i.e. 10°.

When this angle is obtuse (c), **extension** is present. Its range is from 30° to 50°. The margin of individual variations, i.e. 20°, is greater than for flexion.

When these movements become extreme the ankle is not the only active joint: the tarsal joints contribute some range of movement, which is relatively small without being negligible. In extreme flexion (fig. 4), the tarsal joints contribute a few degrees (+) while the plantar arches are flattened. Conversely, during extreme extension (fig. 5) the increase in range (+) is provided by hollowing of the plantar arches.

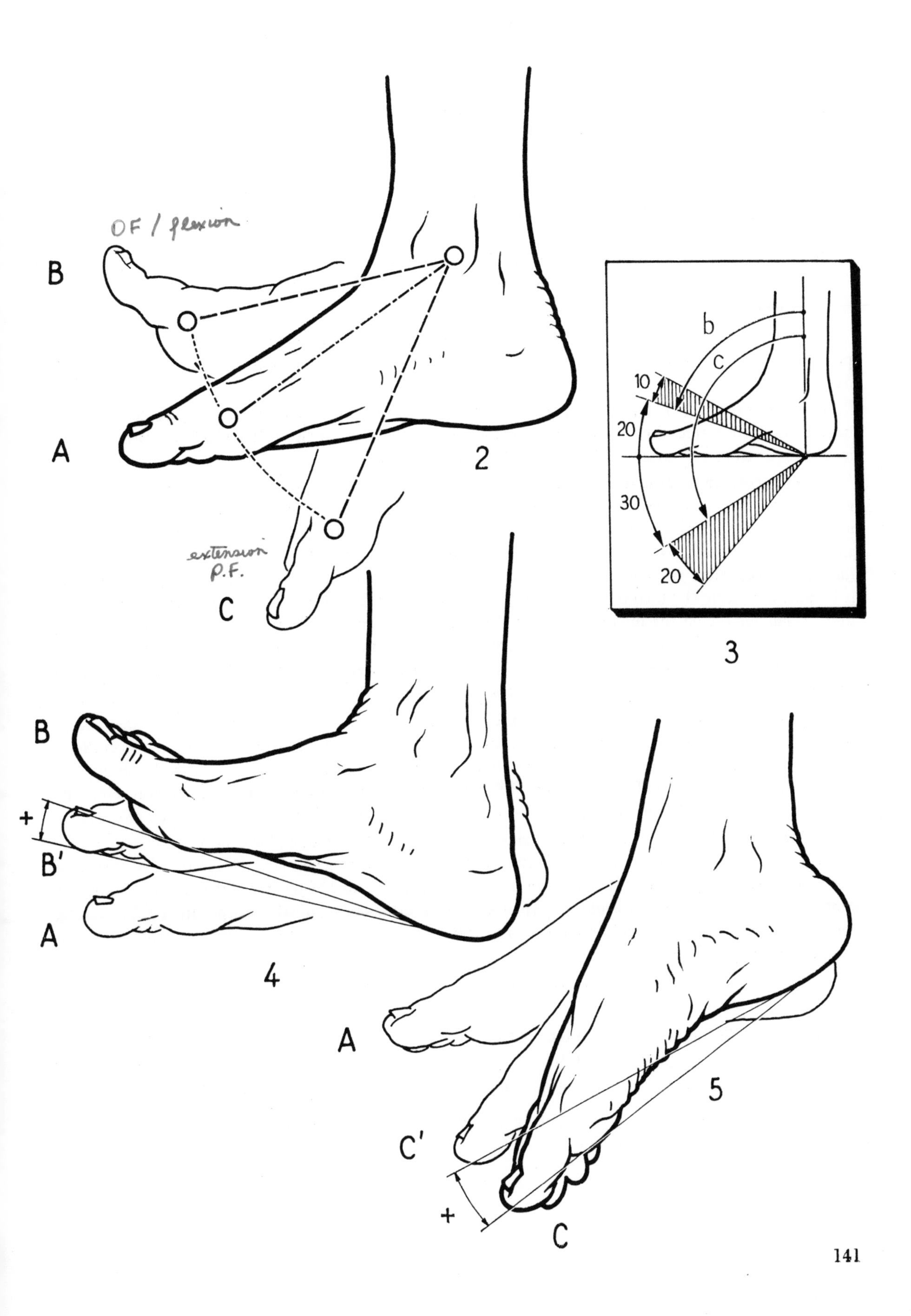

DF / flexion
B
A
2
extension
P.F.
C
b
c
10
20
30
20
3
B
+
B'
A
4
A
5
C'
+
C

THE ARTICULAR SURFACES OF THE ANKLE

(the numbers have the same meaning in all the diagrams.)

If one compares the ankle to a **mechanical model** (fig. 6) it can be described as consisting of:

A lower structure (A), the talus, which bears on its superior aspect a roughly cylindrical surface with its long axis XX′ running transversely.

An upper structure (B), the distal end of the tibia and fibula, forming one structure (shown here as transparent): its lower surface contains a cylindrical cavity corresponding to the cylindrical upper surface of the talus.

The solid cylinder (A), encased within the cylindrical cavity of the upper structure (B), and kept in position by its two flanks, can perform movements of flexion (F) and of extension (E) *around a common axis* XX′.

In the **skeleton** (fig. 7: the ankle opened out and seen from in front and from inside; fig. 8: seen from behind and from inside) the solid cylinder corresponds to the **body of the talus** which has *three surfaces*: a superior or trochlear surface and a medial and a lateral surface.

The **superior or trochlear surface**, convex anteroposteriorly, is depressed centrally by a longitudinal groove—the *groove of the pulley* (1)—bounded by the medial (2) and lateral (3) lip of the pulley. As shown in figure 9 (seen from above), this groove does not strictly lie in a sagittal plane but runs obliquely anteriorly and laterally (arrow Z), along the long axis of the foot as the neck of talus faces anteriorly and medially (arrow T). Therefore the talus is twisted on itself. This diagram also shows that the trochlear surface is broader anteriorly (L) than posteriorly (*l*). This trochlear surface of the talus corresponds to a *reciprocally shaped* surface on the inferior aspect of the tibia (figs. 7 and 8) which is concave anteroposteriorly (fig. 12: sagittal section, viewed from the outside) and has a blunt sagittal ridge (4) to fit into the trochlear groove (fig. 11: frontal section, viewed from in front). One either side of this ridge, a medial (5) and a lateral (6) 'gutter' respectively receive the corresponding lips of the trochlear surface.

The **medial surface** (7) of the body of the talus (fig. 10: seen from inside) is nearly plane, except anteriorly, where it is inclined medially, and lies in a sagittal plane (fig. 9). It articulates with the facet (8) on the lateral surface of the medial malleous (9), which is lined by cartilage continuous with that lining the inferior surface of the tibia. The solid angle (10) lying between these two surfaces of the tibia receives the sharp border (11), which runs between the trochlear surface and the medial surface of the body of the talus.

The **lateral surface** (12) runs obliquely anteriorly and laterally (fig. 8) and is concave supero-inferiorly (fig. 11) as well as anteroposteriorly (fig. 9). It is in contact with the articular facet (13) of the medial surface (fig. 7) of the lateral malleolus (14). This facet is separated from the tibia by the line of the inferior tibiofibular joint (15), padded by a synovial fold (16) (p. 152), which articulates with the sharp border (17) running between the lateral and the trochlear surfaces of the body of the talus. This border is *bevelled* anteriorly (18) and posteriorly (19) (p. 150).

The medial and lateral surfaces of the body of the talus are hemmed in by the two malleoli which are basically different:

The lateral malleolus is bigger than the medial.

It extends farther distally (m, fig. 11).

It lies more posteriorly (fig. 9) so that it runs slightly obliquely (20°) laterally and posterior to the axis XX′.

The third malleolus (fig. 12) is occasionally used descriptively to mean the posterior edge of the lower end of the tibia (20), which sticks out more distally (p) than the anterior margin.

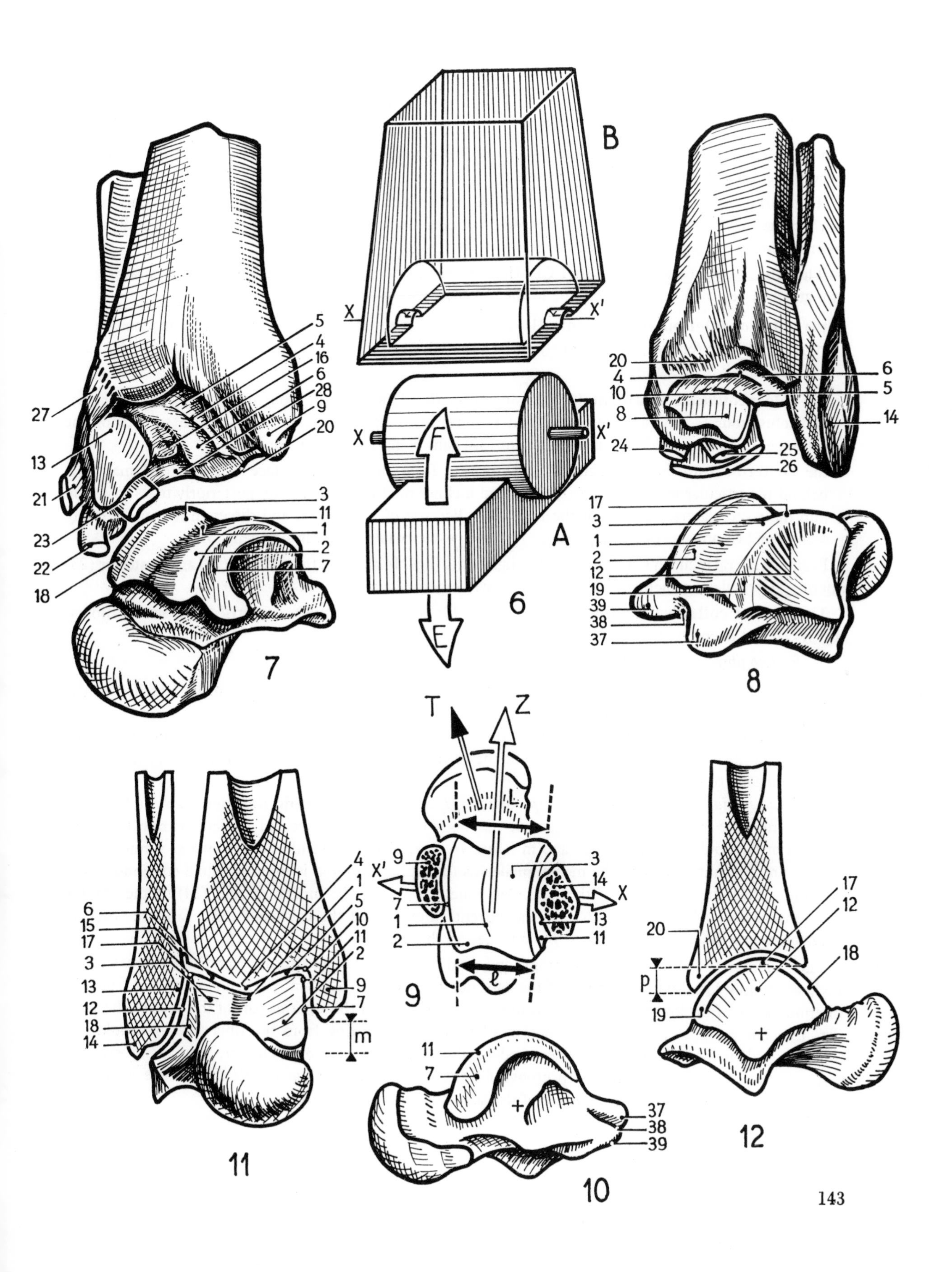

B
X
X'
A
F
E
6
5
4
16
6
28
9
20
27
13
21
23
22
18
3
11
1
2
7
7
20
4
10
8
24
6
5
14
25
26
17
3
1
2
12
19
39
38
37
8
T
Z
L
9
X'
7
1
2
3
14
X
13
11
ℓ
9
6
15
17
3
13
12
18
14
4
1
5
10
11
2
9
7
m
11
7
+
37
38
39
10
20
17
12
18
p
19
+
12

THE LIGAMENTS OF THE ANKLE

(Diagrams are based on Rouvière; the numbers have the same meaning in all diagrams on this page and the preceding page.)

These ligaments consist of two main groups, i.e. the lateral and medial collateral ligaments, and two accessory groups, i.e. the anterior and posterior ligaments.

The **collateral ligaments** form on either side of the joint two powerful fan-like investments, which are attached above at their apices to the corresponding malleolus and which radiate out distally to be inserted into the two posterior tarsal bones.

The **lateral collateral ligament** (LCL) (fig. 13, seen from outside) is made up of three separate bands, two attached below to the talus and one to the calcaneus:

the **anterior talofibular ligament** (21), attached to the anterior margin of the fibular malleolus (14), runs obliquely inferiorly and anteriorly to be inserted into the talus between the lateral articular facet and the mouth of the sinus tarsi;

the **calcaneofibular ligament** (22), arising from the depression in front of the apex of the lateral malleolus, courses obliquely inferiorly and posteriorly to its insertion into the lateral surface of the calcaneus. The lateral talocalcanean ligament (32) runs along its inferior border;

the **posterior talofibular ligament** (23), arising from the medial surface of the lateral malleolus behind the articular facet, runs horizontally and inclines medially and slightly posteriorly to its insertion into the posterolateral tubercle of the talus (37). Because of its position and direction it is more easily seen from behind (fig. 14). It is prolonged by the posterior talocalcanean ligament (31).

From the lateral malleolus spring two other ligaments (figs. 14 and 15): the anterior (27) and posterior (28) inferior tibiofibular ligaments; their significance will emerge later.

The **medial collateral ligament** (MCL) (fig. 16: seen from inside), comprises two sets of fibres, superficial and deep.

The deep fibres consist of two talotibial bands:

the anterior talotibial ligament (25) runs obliquely inferiorly and anteriorly to be attached to the medial aspect of the neck of the calcaneus;

the posterior talotibial ligament (24) runs obliquely inferiorly and posteriorly to be inserted into a deep fossa (fig. 10) on the medial surface of the calcaneus; its most posterior fibres are attached to the posteromedial tubercle (39).

The superficial fibres, triangular in shape, and broad, constitute the **deltoid ligament** (26). In figure 15 (seen from the front) the deltoid ligament has been notched and retracted to demonstrate the deep posterior talofibular ligaments (25) and in figure 16 (seen from the inside) it is shown as a transparent structure. From its origin on the medial malleolus (36) it fans out and is inserted into a continuous line running from the tuberosity of the navicular bone (33), along the medial margin (34) of the plantar calcaneonavicular ligament (p. 162), to the sustentaculum tali of the calcaneus (35). Thus the deltoid ligament, like the lateral ligament, is not attached to the talus.

The **anterior** (fig. 15, seen from in front) and **posterior** (fig. 14, seen from behind) **ligaments** of the ankle are simply localised thickenings of the capsule. The *anterior* ligament (29) runs obliquely from the anterior margin of the lower end of the tibia to the upper surface of the anterior part of the neck of the talus (fig. 13). The *posterior* ligament (30) consists of fibres which spring from the tibia and the fibula and converge to their insertion into the posteromedial tubercle of the talus (39). This tubercle, along with the posterolateral tubercle, forms the deep groove for the flexor hallucis longus (38); this groove is seen to proceed distally along the inferior surface of the sustentaculum tali (p. 182).

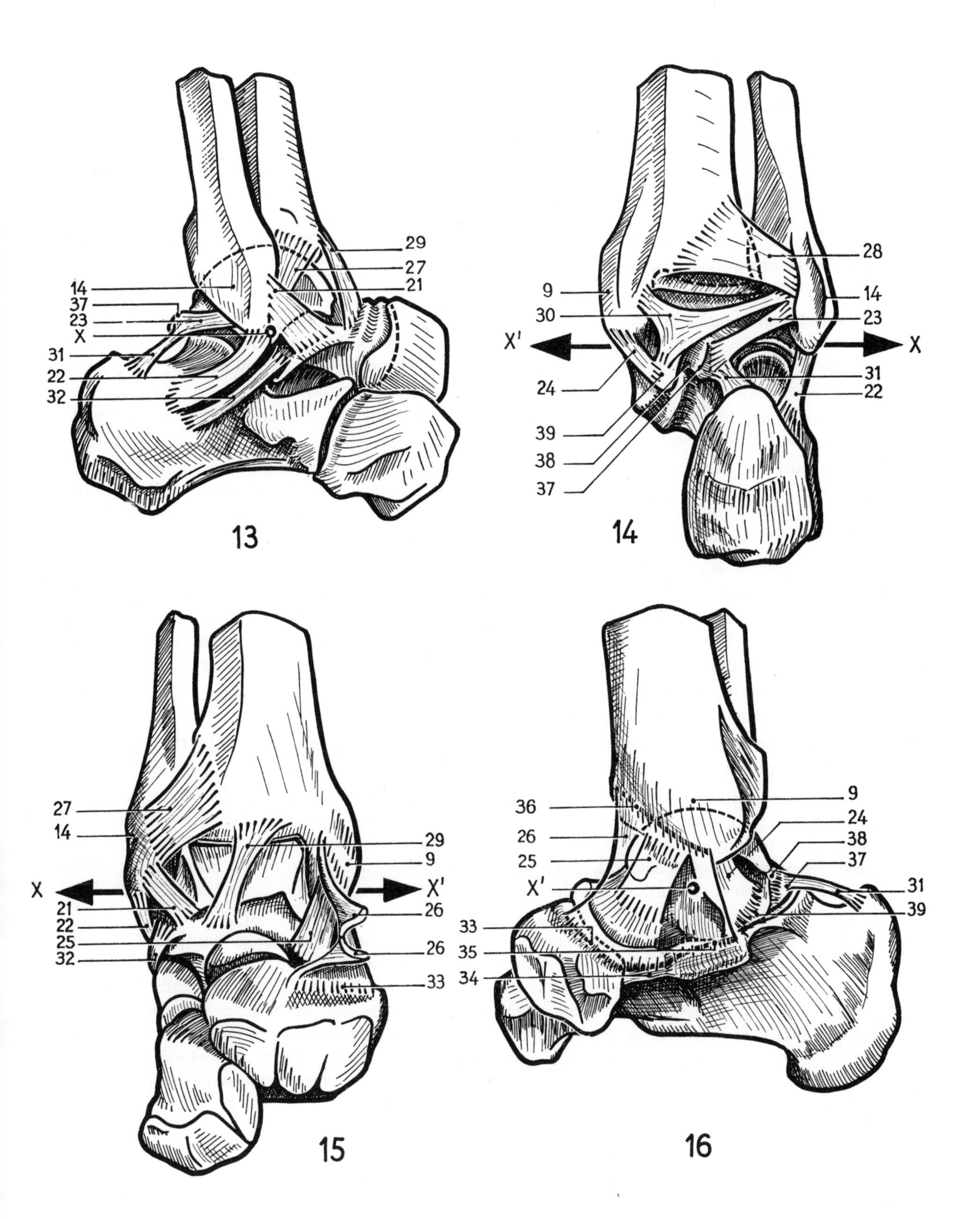
29
27
14
21
37
23
X
31
22
32
13
28
9
14
30
23
X'
X
31
24
22
39
38
37
14
27
14
29
9
X
X'
21
26
22
25
26
32
33
15
9
36
24
26
38
25
37
X'
31
39
33
35
34
16

THE ANTEROPOSTERIOR STABILITY OF THE ANKLE AND THE FACTORS LIMITING FLEXION AND EXTENSION

The range of the movements of flexion and extension is first of all determined by the 'lengths' of the profiles of the articular surfaces (fig. 17). The tibial surface is equivalent to the arc of a circle subtending an angle of 70° at the centre and the trochlear surface of the talus to an arc subtending an angle of 140° to 150°, therefore by simple arithmetic the total range of flexion and extension can be deduced to be 70° to 80°. Since the 'arc length' of the trochlear surface is longer posteriorly than anteriorly it follows that extension has a greater range than flexion.

Flexion is checked (fig. 18) by the following factors:

bony factors: during extreme flexion the upper surface of the neck of the talus comes into contact (1) with the anterior margin of the tibial surface. If flexion continues, the neck of the talus can be fractured. The inferior part of the capsule is prevented from being nipped between the two bones by being pulled up (2) by the flexor muscles, whose sheaths are attached to the capsule;

capsular and ligamentous factors: the posterior part of the capsule is stretched (3) as well as the posterior fibres of the collateral ligaments (4);

one muscular factor: the resistance offered by the tonically active soleus and gastrocnemius muscles (5) usually limits flexion before the other two factors. Hence shortening of these muscles will check flexion prematurely and the ankle may be fixed permanently in a position of extension (talipes equinus); this can be corrected surgically by lengthening the Achilles tendon.

Extension is checked (fig. 19) by similar factors:

bony factors: the posterior tubercles of the talus, especially the posterolateral tubercle, strike (1) the posterior margin of the tibial surface. Occasionally the posterolateral tubercle is fractured during hyperextension but very often this tubercle is separate from the talus and is called the os trigonum. Once more the capsule avoids being nipped (2) by a mechanism similar to that operating during flexion;

capsular and ligamentous factors: the anterior part of the capsule is stretched (3) as well as the anterior fibres of the collateral ligaments (4);

a muscular factor: the resistance offered by the tonically active flexor muscles (5) is the first limiting factor. Hyperactivity of the flexors leads to a flexion deformity of the ankle (talipes calcaneus).

The anteroposterior stability of the ankle and the coaptation of its articular surfaces (fig. 20) depend upon the effect of *gravity* (1), which keeps the talus pressed against the distal surface of the tibia while the *anterior* (2) and *posterior* (3) *margins* of the tibial surface form bony spurs which prevent the talar pulley from escaping anteriorly or posteriorly. The *collateral ligaments* (4) are passively responsible for the coaptation of the surfaces and are assisted by the muscles (not shown here), provided the joint is intact.

When flexion and extension exceed the normal range, one of the limiting factors must give way. Thus, during **hyperextension**, the joint may be dislocated posteriorly (fig. 21) with partial or complete disruption of the capsular ligaments or the posterior margin of the tibial surface (fig. 22) or third malleolus may be fractured with secondary posterior subluxation of the joint. This subluxation tends to recur even after proper surgical reduction, if the 'arc length' of the fractured margin exceeds one-third of the 'arc length' of the tibial surface; fixation by pinning becomes imperative. Likewise, during **hyperflexion**, the joint may be dislocated anteriorly (fig. 23) or there may be fracture of the anterior margin of the tibial surface (fig. 24).

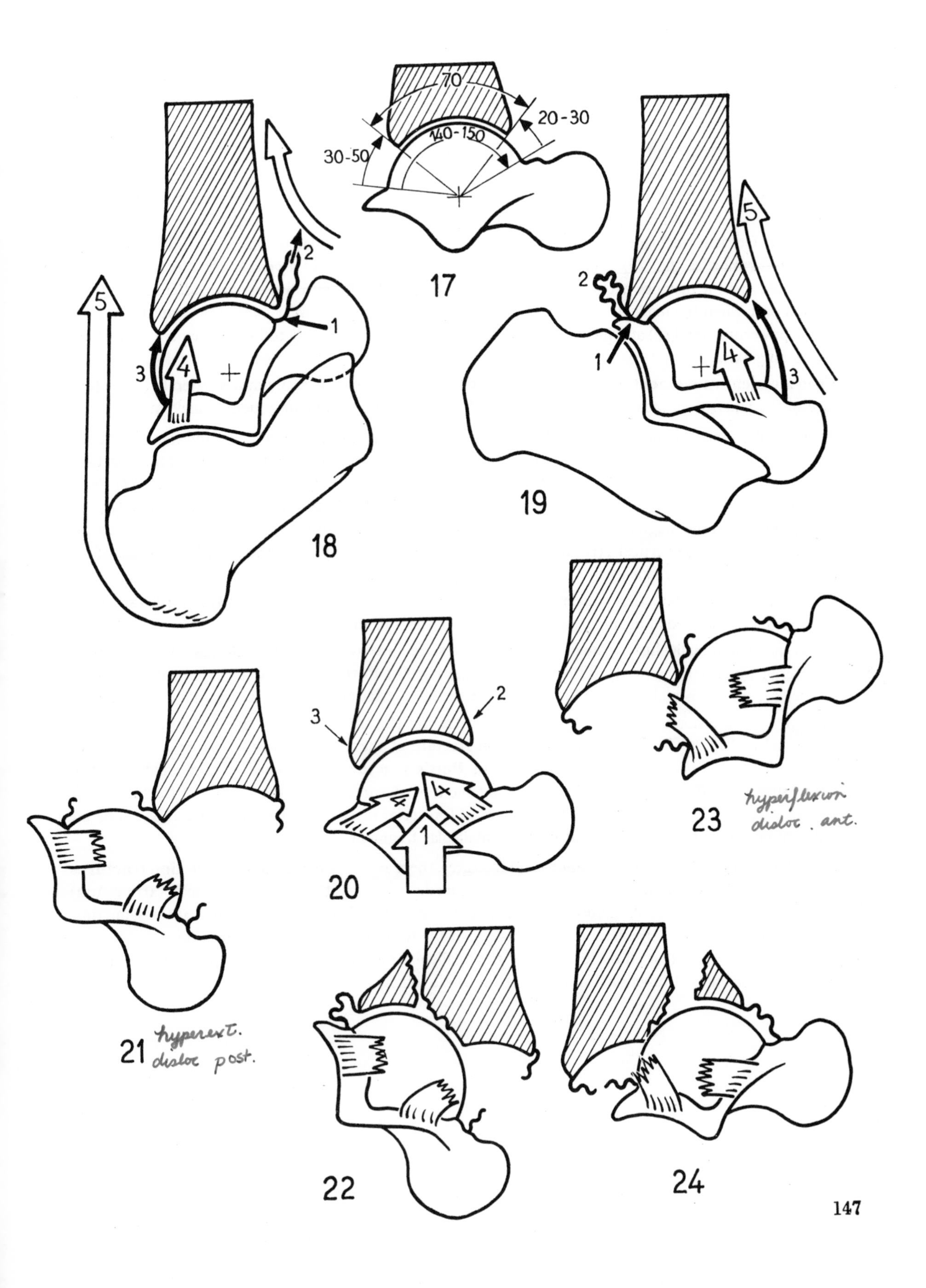
70
20-30
140-150
30-50
17
2
1
5
3
4
18
2
1
5
4
3
19
3
2
4
4
1
20
23
hyperflexion
disloc. ant.
21
hyperext.
disloc post.
22
24

THE TRANSVERSE STABILITY OF THE ANKLE

Being a joint with a single degree of freedom, the ankle cannot, by virtue of its very structure, exhibit movements around its two other axes in space. This stability depends upon the **tight interlocking of its surfaces**: in fact it is analogous to a *tenon-and-mortise joint*, with the talar tenon being tightly fitted into the tibiofibular mortise (fig. 25). The two malleoli, as the two branches of a *pincer*, grip the talus on each side provided that the distance between the lateral (A) and medial (B) malleoli is unchanged. This condition is fulfilled only when the malleoli and the ligaments of the inferior tibiofibular joint (1) are intact. Furthermore the powerful lateral (2) and medial (3) collateral ligaments preclude any rolling movement of the talus about its long axis.

When a **violent movement of abduction** takes place, i.e. the foot is forcibly moved laterally, the lateral surface of the talus strikes against the lateral malleolus and the following consequences may ensue:

the 'malleolar pincer' is disrupted (fig. 26) as a result of rupture of ligaments of the inferior tibiofibular joint (1), this leads to widening of the tibiofibular mortise or **diastasis of the ankle**. Thus the talus is no longer held tightly and can move from side to side (rattling of the talus). It can also (fig. 27) rotate about its long axis (tilting of the talus) and this is made easier if the medial collateral ligament (MCL) is sprained (3) (in the diagram the ligament is shown stretched, i.e. a mild sprain). Finally, the talus can turn (fig. 32) round its vertical axis (arrow Abd) so that the posterior part of the trochlear surface of the talus fractures the *posterior margin* of the tibial surface (arrow 2);

if this movement of abduction continues (fig. 31) the medial collateral ligament (3) is torn: this is the *severe sprain* of the medial ligament associated with diastasis of the ankle;

or else the medial malleolus (B) snaps (fig. 29) at the same time as the lateral malleolus (A) snaps above the inferior tibiofibular joint (1). This represents one form of **Pott's fracture**. Occasionally the fibular fracture occurs much higher at the level of the neck: this is Maisonneuve's fracture (not shown here);

very often the inferior tibiofibular ligaments resist tearing (fig. 28), especially the anterior. Fracture of the medial malleolus (B) is then associated with fracture of the lateral malleolus *before or through the inferior tibiofibular joint*. This is another form of **Pott's fracture**. Occasionally the medial malleolus fails to snap (fig. 30) and the medial collateral ligament is ruptured (3). In these types of fracture a chip of bone is frequently broken off the 'third malleolus' (posterior margin of the tibia): this can be a separate fragment or it may form a single unit with the malleolar fragment.

In addition to these abduction dislocations and fractures, there are also **bimalleolar adduction fractures** (fig. 33). As the foot is adducted the talus (fig. 32) is made to rotate about its vertical axis (arrow Add) and its medial surface breaks off (arrow 3) the medial malleolus (B). The talus is also tilted at the same time and this leads to fracture of the lateral malleolus (A) at the level of the tibial articular surface.

It goes without saying that all these lesions of the 'malleolar pincer' require proper treatment if one is to restore the structural and functional integrity of the ankle joint.

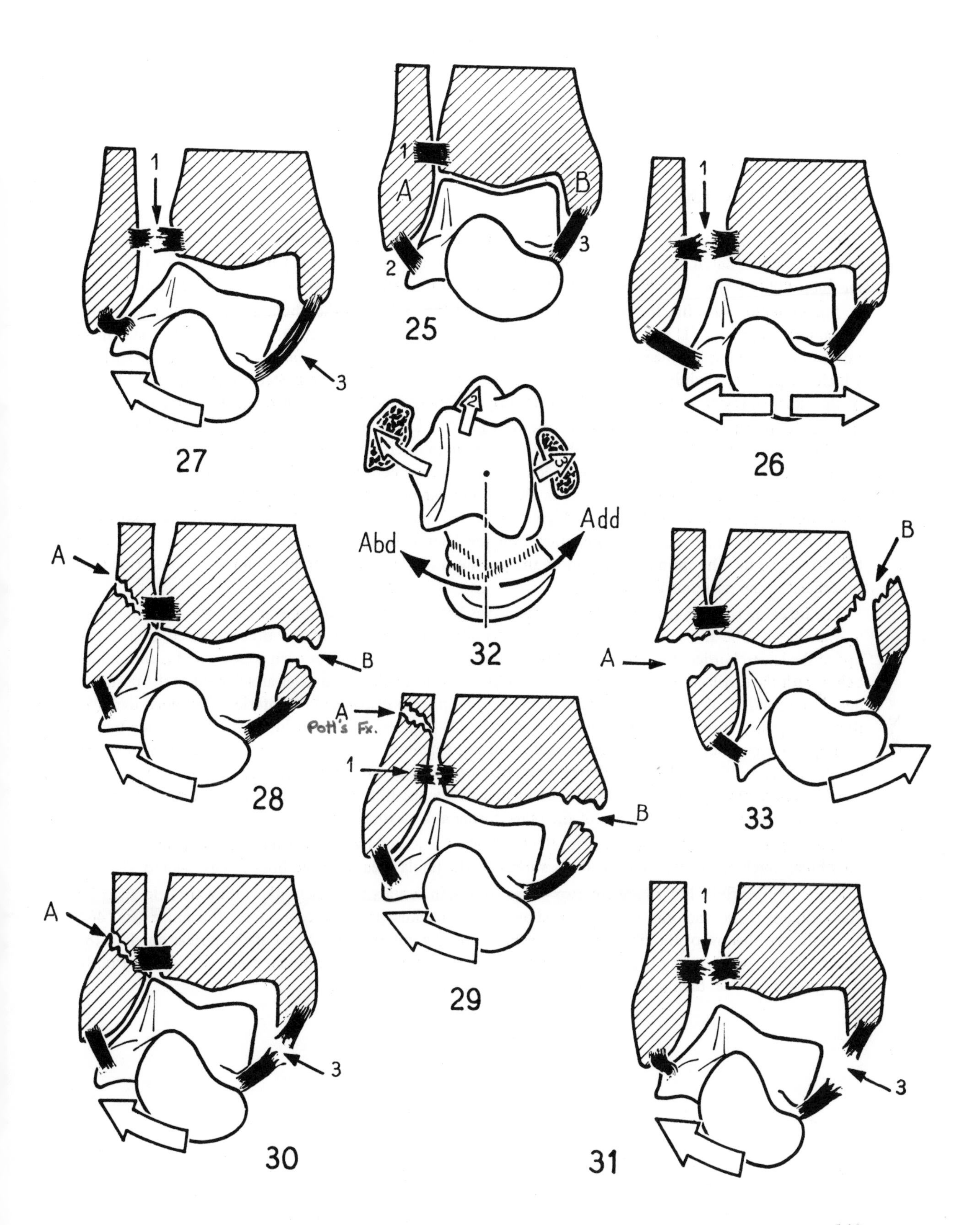
1
A
B
2
3
25
1
3
27
1
26
Abd
Add
32
A
B
28
B
A
33
A
Pott's Fx.
1
B
29
A
3
30
1
3
31

THE TIBIOFIBULAR JOINTS

The tibia articulates with the fibula at its two extremities, i.e. at the superior tibiofibular joint (figs. 34 to 36) and the inferior tibiofibular joint (figs. 37 to 39). It will be shown in the next page that these two joints are *mechanically linked to the ankle* and it is therefore logical to study these two joints in relation to the ankle.

The **superior tibiofibular joint** is clearly exposed (fig. 34) when the fibula is rotated after sectioning the anterior ligament (1) and the anterior expansion (2) of the biceps tendon (3). The joint then opens out around the hinge formed by its posterior ligament (4). It is a *plane joint* with oval articular surfaces which are plane or slightly convex. The tibial articular facet (5) lies on the posterolateral aspect of the rim of the tibial condyle; it faces obliquely posteriorly, inferiorly and laterally (arrow). The fibular facet (5) lies on the upper surface of the head of the fibula and it looks anteriorly, superiorly and medially. It is overhung by the styloid process of the fibula (7) which gives insertion to the tendon of the biceps femoris (3). The lateral collateral ligament of the knee joint (8) is attached between the biceps insertion and the fibular facet. Figure 35 (seen from behind) shows clearly how far posteriorly the fibular head lies; it also shows the anterior ligament of the joint (1), which is short and quadrilateral, and the thick tendinous expansion of the biceps (2) which runs to its insertion into the lateral condyle of the tibia. Figure 36 (seen from behind) illustrates the intimate relation of the popliteus (9) with the superior tibiofibular joint as it runs superficial to the posterior ligament (4).

The **inferior tibiofibular joint** (fig. 37: opened as before) contains no articular cartilage and is therefore a *syndesmosis*. The tibial facet (1) is the rough concave fibular notch of the tibia bounded by the two lips of the lateral border of the tibia. The fibular facet (2) is convex, plane or even concave and is continuous below with the cartilage-lined fibular articular facet (3) of the ankle which gives attachment to the posterior talofibular band (4) of the lateral collateral ligament. The *anterior ligament* of the inferior tibiofibular joint (5), thick and pearly, runs obliquely inferiorly and laterally (fig. 38, seen from in front). Its inferior border overlaps the tibiofibular mortise laterally and so during flexion of the ankle it 'nicks' the lateral ridge of the trochlear surface of the talus (double arrow). The *posterior ligament* (6), thicker and broader (fig. 39: seen from behind), runs a long way towards the medial malleolus; likewise it 'nicks' the posterior part of the lateral ridge of the trochlear surface of the talus during ankle extension.

In addition to the ligament of the joint, the two bones are joined by the **interosseous ligament** between the fibular notch of the tibia and the inner surface of the fibula (heavy dotted lines in figs. 34 and 37).

In the inferior tibiofibular joint the two bones are not in contact with each other: they are held apart by fibro-adipose tissue and this gap can be shown on skiagrams properly centred on the ankle (fig. 40). Normally the shadow of the fibula (c) encroaches upon the anterior border (a) of the fibular notch of the tibia by 8 mm., which it is only 2 mm. away from the posterior border of the fibular notch (b). If the distance cb is greater than ac, then *diastasis of the ankle joint* is said to be present.

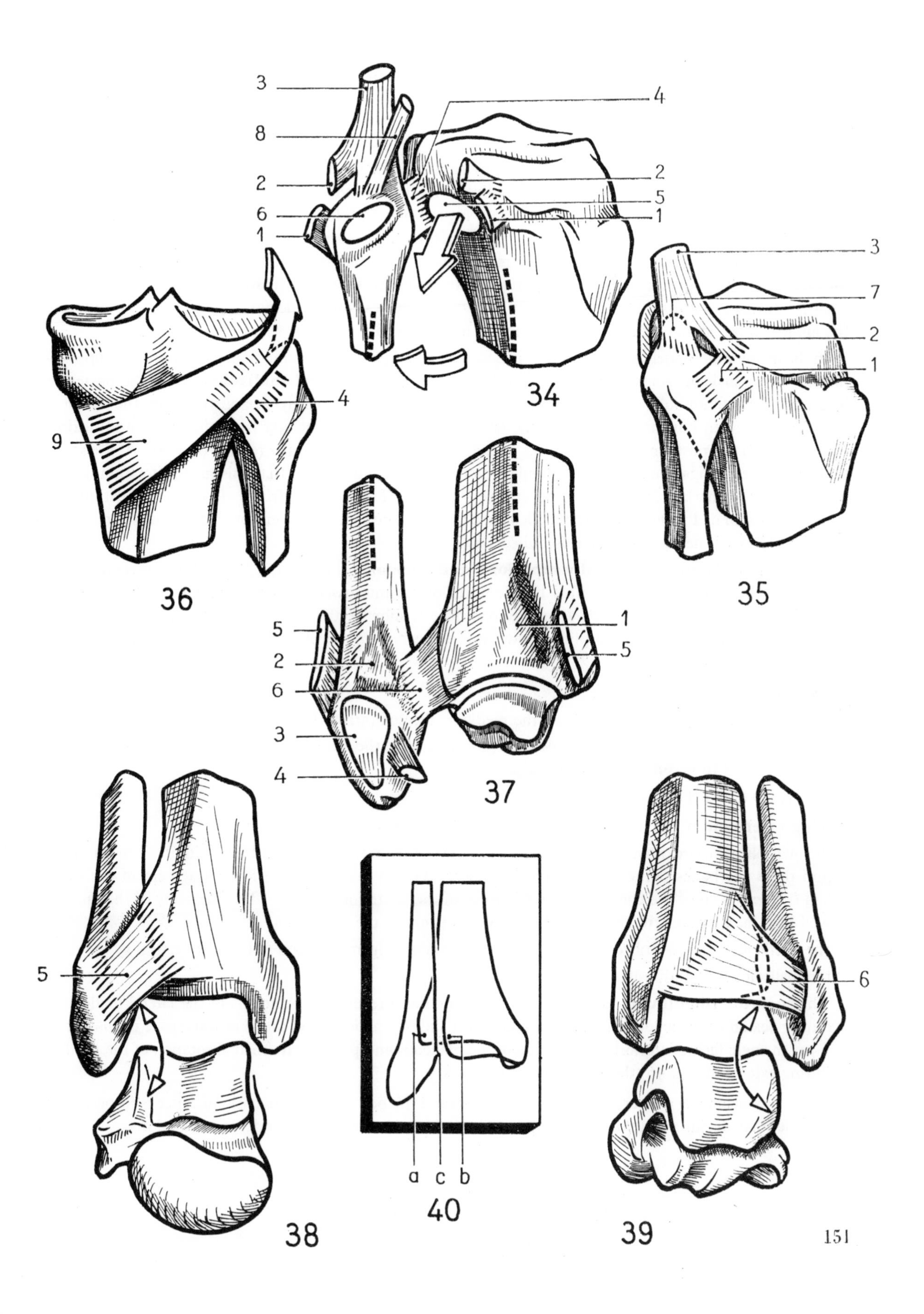
3
8
2
6
1
4
2
5
1
34
3
7
2
1
4
9
36
35
5
2
6
3
4
1
5
37
5
6
a
c
b
40
38
39

THE PHYSIOLOGICAL FUNCTIONS OF THE TIBIOFIBULAR JOINT

Flexion and extension of the ankle automatically call into action the two tibiofibular joints which are therefore **mechanically linked** to the ankle.

The **inferior tibiofibular joint** is the first to be recruited. Its mode of action has been well worked out by Pol le Coeur (1938) and depends essentially on the shape of the trochlear surface of the talus (fig. 41: seen from above). Its medial surface (M) lies in a sagittal plane while the lateral surface (L) lies in a plane which runs obliquely anteriorly and laterally. Therefore the width of the trochlear surface is smaller (posteriorly (aa′) than anteriorly (bb′)) by 5 mm. Hence, if the medial and lateral surfaces of the body of the talus are to be gripped tightly the **intermalleolar space must vary within certain limits**, i.e. being smallest during extension (fig. 42: seen from below) and greatest during flexion (fig. 43). On the cadaver, the ankle can be extended simply by pressing the malleoli firmly together.

In the anatomical model (figs. 42 and 43) it is also obvious that this movement of separation and approximation of the malleoli is associated with **axial rotation of the lateral malleolus**, while the posterior ligament of the tibiofibular joint (1) acts as a hinge. This rotation is easily demonstrated by the use of a pin which is driven horizontally through the lateral malleolus. When moving from the position of flexion (mm′., fig. 43) to a position of extension (nn′, fig. 42) the malleolus is medially rotated over 30°. At the same time the posterior ligament of the tibiofibular joint (2) is stretched, Note that this medial rotation of the malleolus is less marked in life but is nevertheless present. The synovial fringe (f) contained within the joint is displaced as follows: when the malleoli are approximated during extension (fig. 44) it is forced out distally (1); during flexion (fig. 45) it is pulled up (2).

Finally the fibula **moves vertically superiorly and inferiorly** (figs. 46 and 47: the fibula is represented by a ruler). Being attached to the tibia by the fibres of the interosseous membrane which run obliquely inferiorly and laterally (for clarity's sake only one fibre is shown), the fibula is lifted slightly as it moves away from the tibia (fig. 47) and is pulled down as it draws near to the tibia (fig. 46). To sum up the movements of the fibula:

During flexion of the ankle (fig. 48):

the lateral malleolus moves *away* from the medial malleolus (arrow 1).

at the same time it is slightly *pulled superiorly* (arrow 2) while the fibres of the tibiofibular and interosseous ligaments tend to become horizontal (xx′);

finally, the fibula is *medially rotated* (arrow 3).

During extension of the ankle (fig. 49) the converse takes place:

the malleoli are approximated *actively* (arrow 1): contraction of the posterior tibialis with its fibres inserted into these two bones tightens the 'malleolar pincer' (fig. 50: section of the lower fragment of the right leg; the arrows show the contraction of the fibres of the posterior tibialis). Thus the body of the talus is well held in, whatever the degree of flexion or extension of the ankle;

the lateral malleolus is pulled *inferiorly* (arrow 2) while the ligaments tend to become vertieal yy′;

the malleolus is slightly *rotated medially* (arrow 3).

The **superior tibiofibular joint** is called into action as a result of movements of the lateral malleolus: during flexion of the ankle (fig. 47) the fibular facet slides superiorly and the joint interspace opens out to form an angle facing inferiorly (separation of the malleoli) and posteriorly (medial rotation of the fibula); during extension (fig. 46) the exact converse occurs.

These displacements are very small but they occur, and the best proof of their significance is to be found in the fact that during evolution the superior tibiofibular joint has not yet undergone ankylosis.

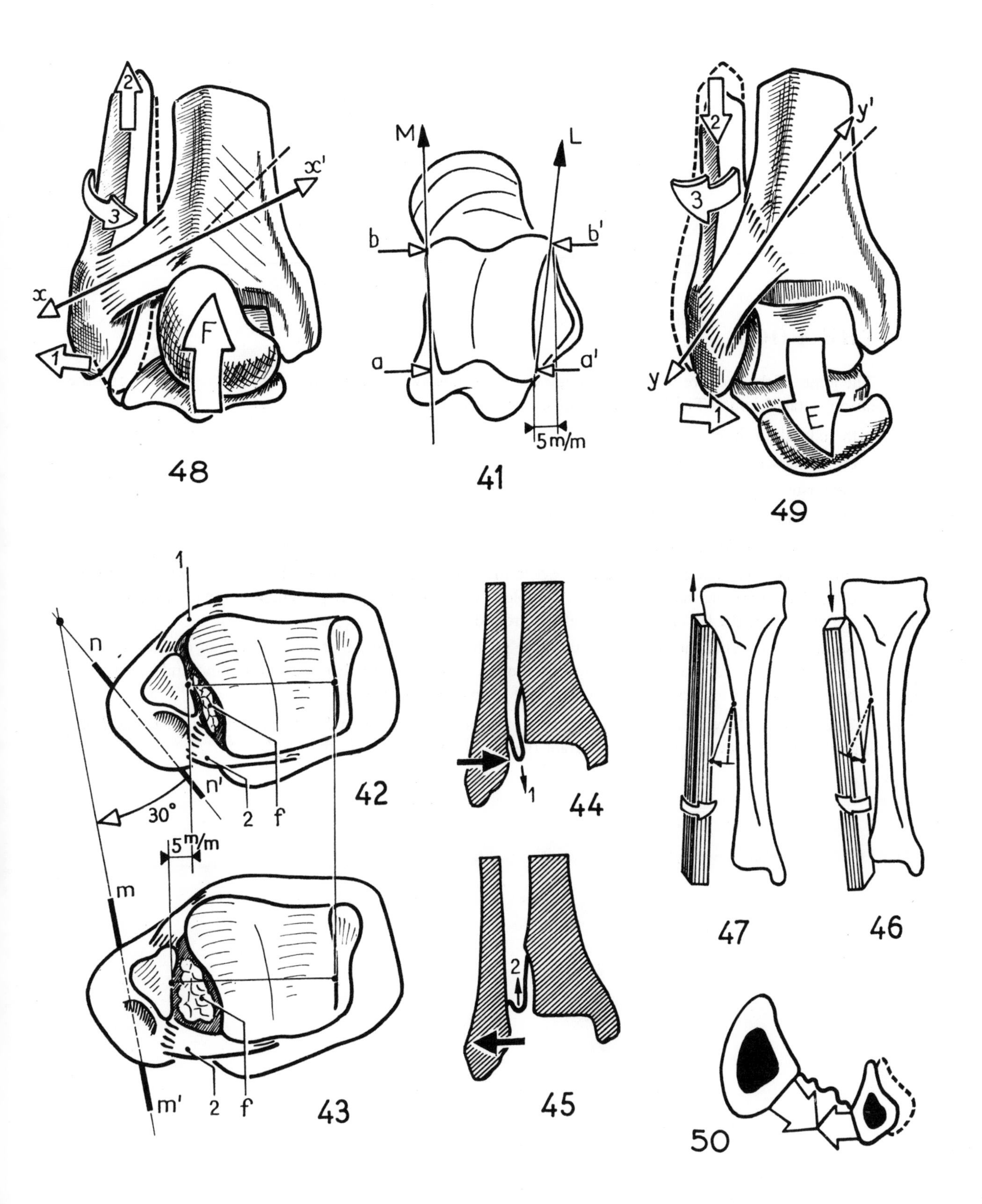

2
3
x'
x
F
1
48
M
L
b
b'
a
a'
5 m/m
41
2
y'
3
y
1
E
49
1
n
n'
30°
2
f
42
5 m/m
m
m'
2
f
43
1
44
2
45
47
46
50

THE FOOT

The joints of the foot are many and complex and fall into two main groups: the intertarsal joints and the tarsometatarsal joints. The important joints are the following:

the talocalcanean or subtalar joint;

the midtarsal or transverse tarsal joint;

the tarsometatarsal joint;

the cubonavicular joint and the cuneonavicular joint.

These joints perform a *dual function.*

Firstly, they orientate the foot with respect to the other two axes in space (the ankle controls movements of the foot in the sagittal plane) so that the sole of the foot is correctly presented to the ground, whatever the position of the leg and the slope of the ground.

Secondly, they alter the shape and curvature of the arches of the foot so that the foot can adapt to the irregularities of the ground; they thus interpose a shock-absorber between the ground and the weight-bearing foot and impart greater elasticity and suppleness to the step.

These joints, therefore, play a vital part in the foot. On the other hand the metatarsophalangeal and interphalangeal joints are far less important than their counterparts in the hand.

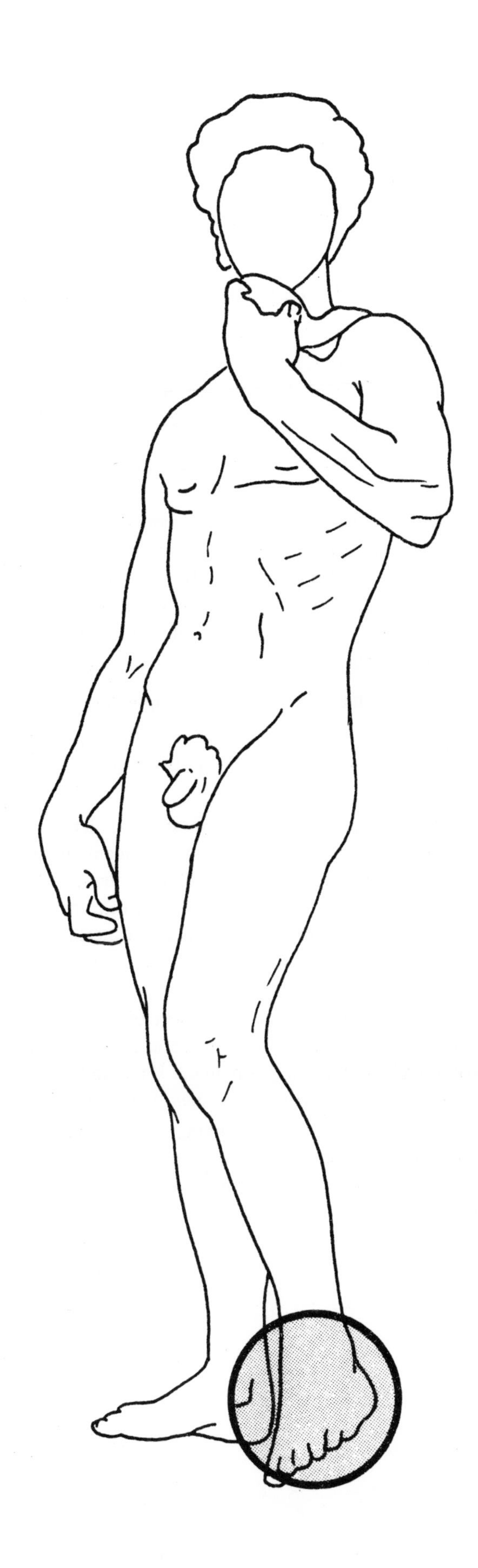

THE MOVEMENTS OF LONGITUDINAL ROTATION AND THE SIDE-TO-SIDE MOVEMENTS OF THE FOOT

In addition to movements of flexion and extension, which occur at the ankle, the foot can move about the vertical axis of the leg (axis Y, p. 138) and about its own horizontal and longitudinal axis (axis Z).

About the vertical axis Y occur movements of **adduction** and **abduction**:

adduction (fig. 2): when the tips of the toes move towards the plane of symmetry of the body and face inwards;

abduction (fig. 3): when the tips of the toes move away from the plane of symmetry and point outwards.

The total range of these movements, when they occur exclusively in the foot, is from 35° to 45° (Roud). However, these movements of the tips of the toes in the horizontal plane can also be achieved by lateral or medial rotation of the leg (knee flexed) or by rotation of the whole lower limb from the hip (knee extended). They have then a much greater range, attaining a maximum of 90° each way in ballerinas.

About the longitudinal axis Z the foot can turn so that the sole of the foot can face either:

medially (fig. 4): by analogy with the upper limb this is *supination*;

laterally (fig. 5): *pronation*.

The range of supination is 52° (Biesalski and Mayer, 1916) and is greater than that of pronation (25° to 30°).

These movements of adduction and abduction and rotation, as defined, do not in life occur *exclusively in the joints of the foot*. In fact, it will be shown later that movements in any one of the planes must needs be associated with movements in the other two planes. Thus adduction is necessarily accompanied (figs. 2 and 4) by supination and a slight measure of extension. These three component movements are characteristic of the position known as *inversion*. If the extension component is cancelled by flexion of the ankle the position of the foot is known as *talipes varus*. Finally if lateral rotation at the knee compensates for adduction of the foot, then the movement of *apparently pure supination* is produced.

Conversely (figs. 3 and 5), abduction is necessarily associated with pronation and flexion: this leads to *eversion*. If the flexion component is cancelled by extension of the ankle (in the diagrams it is overcompensated in extension) the position of *talipes valgus* is obtained. If in addition medial rotation of the knee makes up for abduction of the foot, then a movement of *apparently pure pronation* is achieved.

Thus, barring any compensatory movements occurring at joints outside the foot, adduction can never be associated with pronation and, vice versa, abduction can never be associated with supination. Therefore there are in the foot *forbidden combinations* of movement resulting from the very structure of its joints.

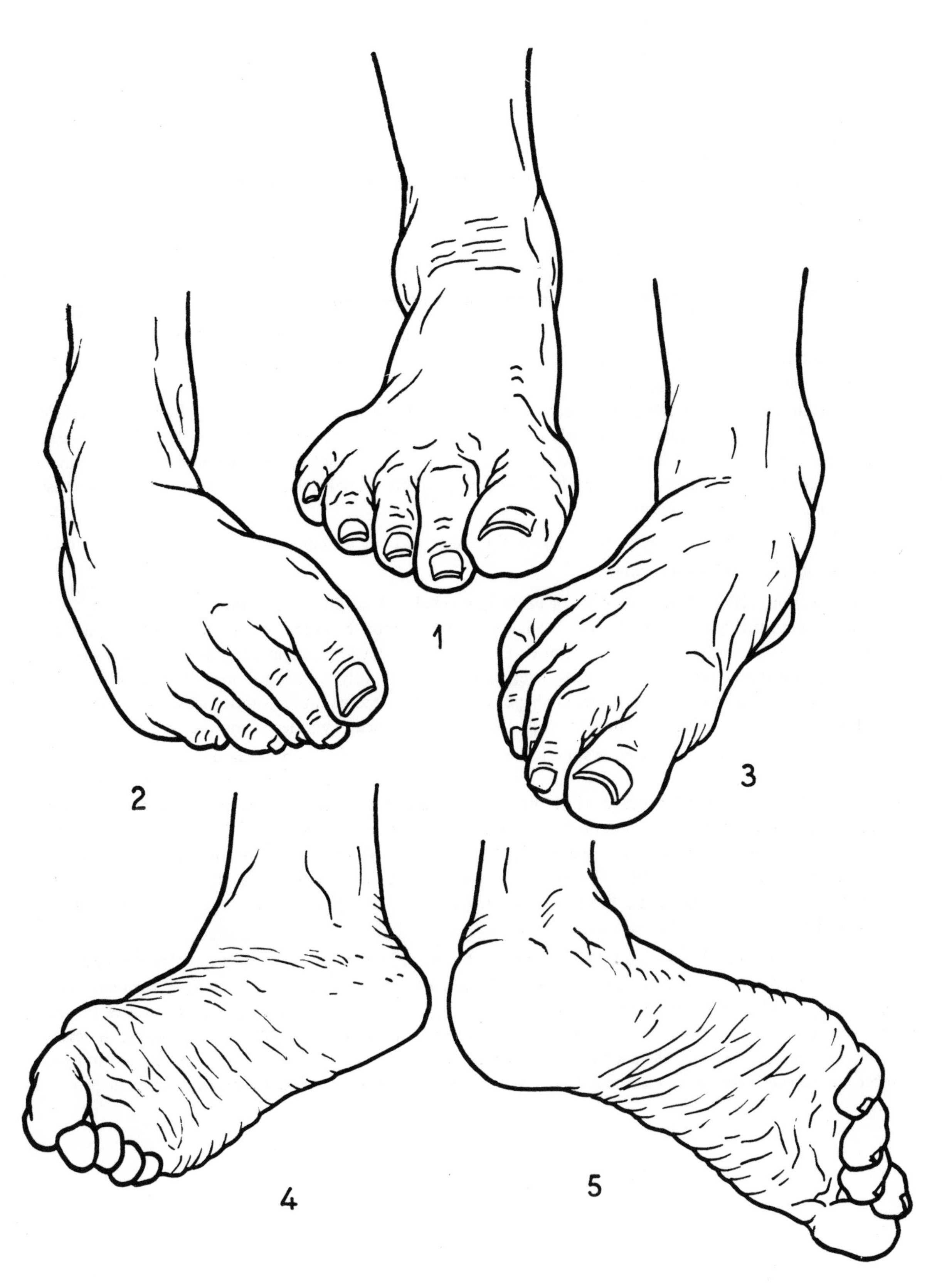
1
2
3
4
5

SUBTALAR (TALOCALCANEAN) JOINT: THE ARTICULAR SURFACES

(the numbers have the same meaning in all the diagrams)

The inferior surface of the talus articulates (A, fig. 6: the bones have been separated and the talus rotated around its hinge-like axis *xx'*) with the superior surface of the calcaneus (B, fig. 6). This occurs at two separate articular facets, which constitute together **the subtalar joint**:

The posterior surface of the talus (a) is in contact with the broad (a′) superior surface of the calcaneus (also known as the thalamus of Destot). These two surfaces are united by ligaments and a capsule, so that the joint is anatomically distinct.

The small surface (b) on the inferior surface of the neck and head of the talus rests on the anterior surface of the calcaneus (b′), which is obliquely set and is supported by the sustentaculum tali and the neck of the calcaneus. The joint, however, also includes the posterior surface of the navicular bone (d′) which articulates with the head of the talus (d). This joint, properly the talocalcaneonavicular joint, is the most medial of the mid-tarsal joints.

Before studying the function of these joints, the shape of their articular surfaces must be understood. These joints are of the **plane variety**:

The superior surface of the calcaneus (a′) is roughly oval with its great axis running anterolaterally; it is convex about this great axis and plane or slightly concave about the other axis (fig. 7: seen from outside; fig. 8: seen from inside). Therefore it can be viewed as analogous to a segment of a cylinder (f) with its axis running obliquely postero-anteriorly, lateromedially and slightly supero-inferiorly. The corresponding talar surface (a) also has this cylindrical shape with a similar radius and a similar axis, except that the surface of the talar cylinder is concave while that of the calcanean cylinder is convex.

As a whole the head of the talus is spherical and the bevelled surfaces on its circumference can be considered as facets chiselled out on a sphere (broken line) with centre g (fig. 6). Thus the anterior surface of the calcaneum (b) is biconcave while the corresponding talar surface is reciprocally biconvex. Very often the calcanean surface is indented in its middle and assumes the shape of the sole of a shoe; occasionally it is subdivided into two separate facets (figs. 7 and 8), one resting on the neck (b'_1) and the other on the sustentaculum tali (b'_2). It has been noted that the stability of the calcaneus is a function of the surface area of the latter facet. Occasionally the talus also presents two separate articular facets (b_1 and b_2).

The calcanean facet (b′ or $b'_1 + b'_2$) is itself part of a much longer spherical surface which comprises in addition the posterior surface of the navicular (d)′ and the upper edge of the plantar calcaneonavicular ligament (c′). With the help of the deltoid ligament (5) and the capsular ligament these surfaces form a spherical cavity which receives the head of the talus. On the head of the talus corresponding articular facets are present: the bulk of its articular surface (d) lodges into the navicular; between this (d) and the facet for the calcaneus (b) lies a triangular facet (c), which receives the calcaneonavicular ligament (c′).

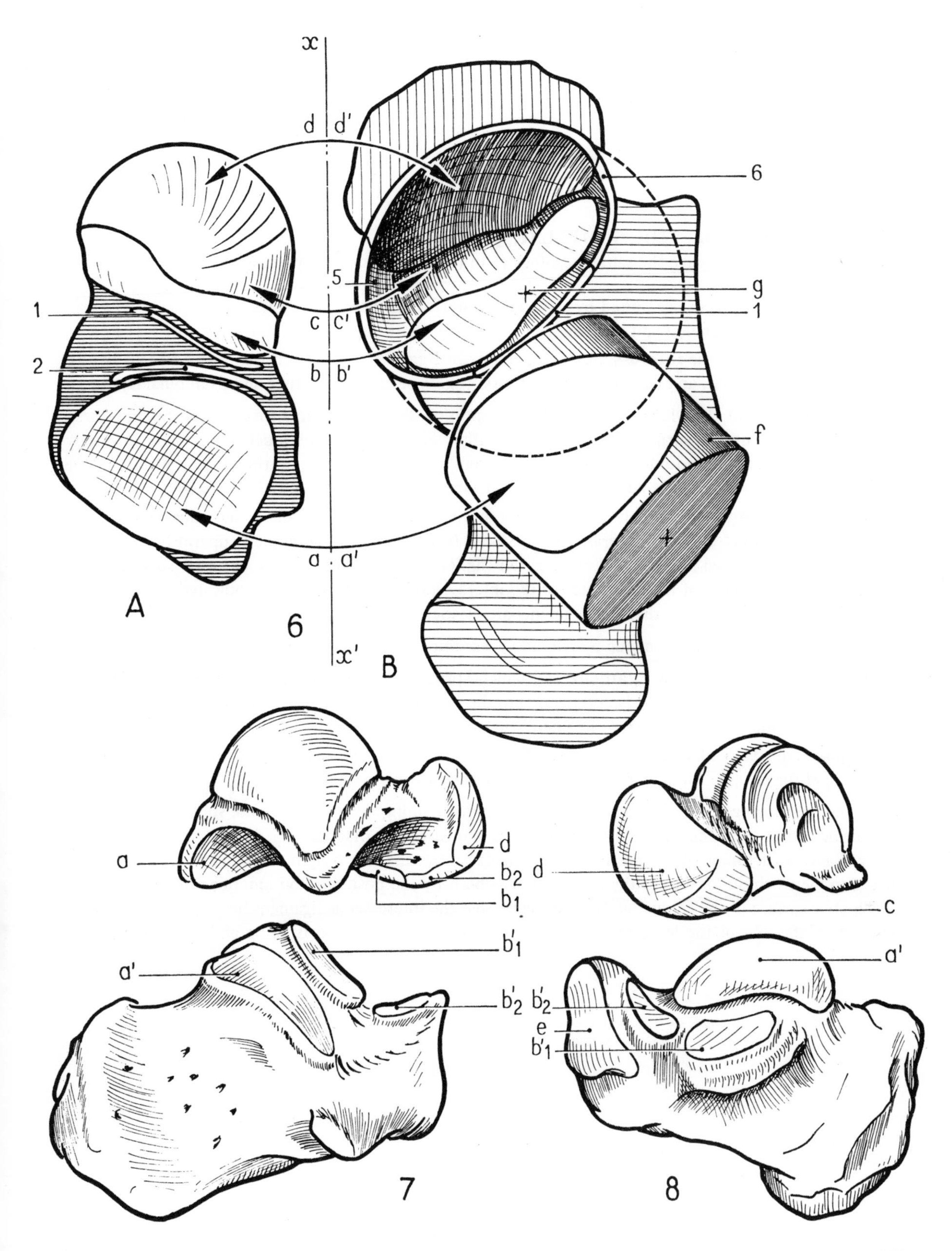

x
d
d'
6
5
g
1
c
c'
1
2
b
b'
f
a
a'
A
6
x'
B
a
d
b2
b1
d
c
b'1
a'
a'
b'2
b'2
e
b'1
7
8

THE LIGAMENTS OF THE SUBTALAR JOINT

(the numbers and letters have the same meanings as in the diagrams of the previous page).

The talus and the calcaneus are united by short and powerful ligaments since they are subjected to severe stress during walking, running and jumping.

The main ligament is the **interosseous talocalcanean ligament** which consists of two fibrous bands, thickset and quadrilateral, and occupies the sinus tarsi (fig. 9: seen from in front and from outside):

The anterior band (1), which is attached to the depths of the sinus calcanei (the floor of the sinus tarsi) just behind the anterior surface. Its dense and pearly white fibres run obliquely superiorly, anteriorly and laterally to be inserted into the sinus tali (the roof of the sinus tarsi), on the inferior surface of the talar neck and just behind the edge of the articular facet of the head (fig 6, A).

The posterior band (2) lies behind the former and is attached to the floor of the sinus tarsi i.e. the sinus calcanei just in front of its superior articular facet. Its thick fibres run obliquely superiorly, posteriorly and laterally to be fixed into the roof of sinus tali (fig. 6, A), just in front of the posterior surface of the talus. The arrangement of the two bands of the interosseous ligament is clearly indicated by separating the two bones, i.e. assuming that the ligaments are elastic (fig. 10).

The talus is united to the calcaneus by two other less important ligaments (figs. 9 and 10):

The lateral talocalcanean ligament (3), which springs from the lateral tubercle of the talus, runs obliquely inferiorly and posteriorly and parallel to the intermediate band of the lateral collateral ligament of the ankle joint and is attached to the lateral surface of the calcaneus.

The posterior talocalcanean ligament (4) is a thin fibrous band running from the posterolateral tubercle of the talus to the upper surface of the calcaneus.

The interosseous ligament plays an essential part in maintaining the stability of the subtalar joint at rest and during activity. It occupies a central position as shown by the diagram (fig. 11), where a transparent pulley-shaped structure is placed on the calcaneum. It is clear that the weight of the body, transmitted by the leg to the trochlear surface of the talus, is borne both by the posterior and anterior articular facets of the superior surface of the calcaneus. It is also clear that the interosseous ligament lies exactly on the prolongation of the axis of the leg (cross in circle) so that it is subjected to twisting and stretching strains (p. 164).

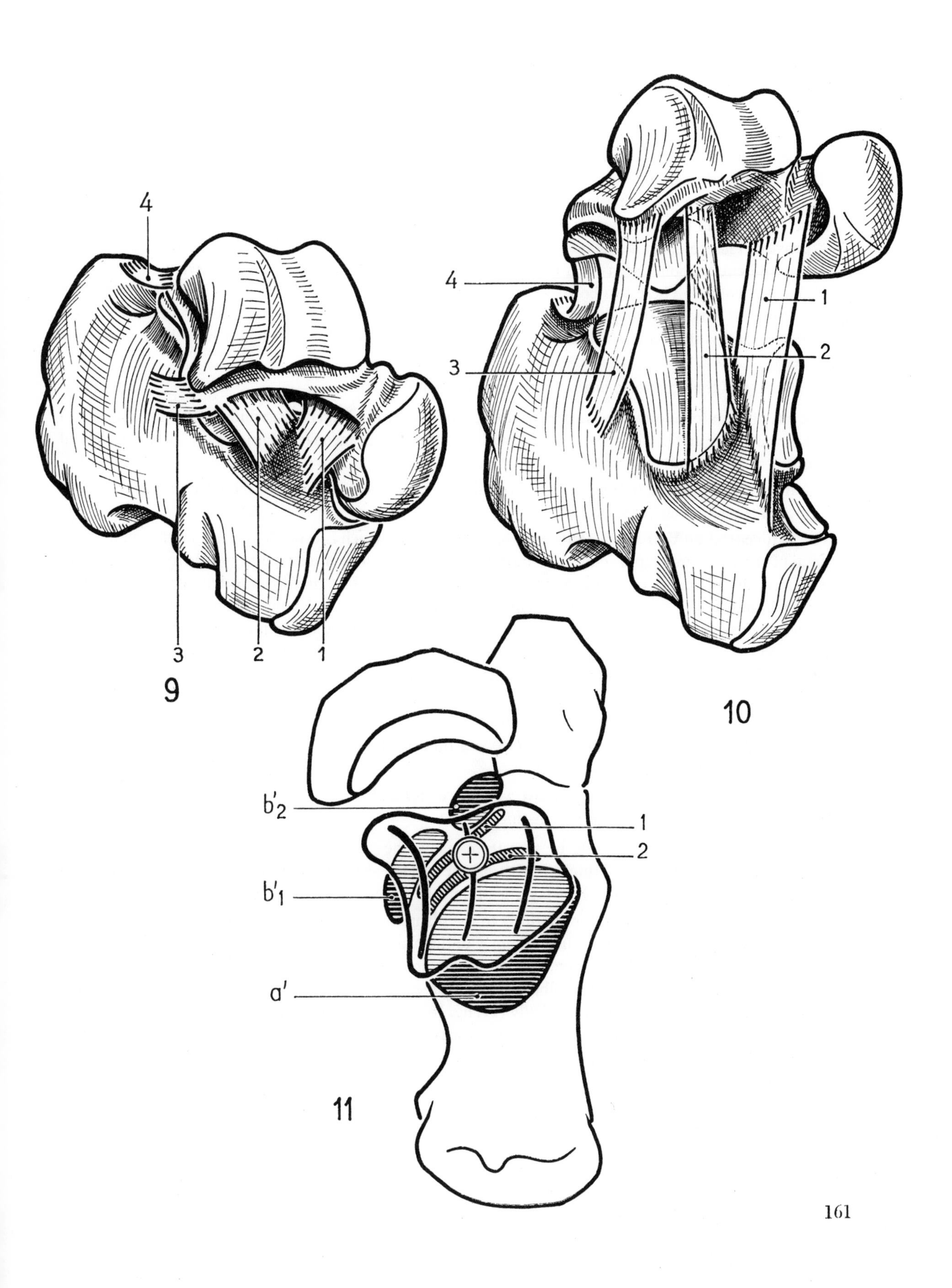

4
3
2
1
9
4
3
1
2
10
b'2
1
2
b'1
a'
11

THE TRANSVERSE TARSAL (MIDTARSAL) JOINT

(the numbers and letters have the same meaning as in the diagrams of the two preceding pages).

When the transverse tarsal joint is opened and the navicular and cuboid bones are tilted distally (fig. 12, according to Rouvière), it is seen to consist of two joints: medially, the talonavicular joint concave posteriorly (p. 158); laterally, the calcaneocuboid joint slightly concave anteriorly, so that the midtarsal line is S-shaped when seen from above. The anterior surface (e) of the calcaneum has a complex shape: transversely, it is concave in its upper part and convex in its lower part; vertically (supero-inferiorly) it is first concave and then convex. The posterior surface of the cuboid has the corresponding shape but often (fig. 17: the navicular and cuboid bones seen from behind) it is extended by a facet (e'_2) on the navicular, which lies medial to the cuboid. These two bones articulate with each other (the cubonavicular joint) by two plane surfaces (h and h′) and are strongly united by three ligaments, a lateral dorsal ligament (5)' a medial plantar ligament (6) and a short and thick interosseous ligament (7) (here the two bones have been artificially separated).

The ligaments of the transverse tarsal joint are five in number:

The plantar calcaneonavicular ligament (c′) or the 'spring' ligament, which unites the calcaneum and the navicular and participates in the formation of the talocalcaneonavicular joint (p. 158). The deltoid ligament (p. 144) is attached to its medial border (8).

The dorsal talonavicular ligament (9), running from the dorsal surface of the talar neck to the dorsal surface of the navicular bone (fig. 16).

The bifurcated or Y-shaped ligament (figs. 13 and 16), which is centrally placed and forms the key-stone of the transverse tarsal joint. It comprises two bands which arise together (10) from the anterior part of the dorsal surface of sustentaculum tali of the calcaneus. **The medial band** or lateral calcaneonavicular ligament (11) lies in a vertical plane and is inserted into the lateral extremity of the navicular; its deep border occasionally unites with the plantar calcaneonavicular ligament so that the transverse tarsal joint is split into two separate synovial cavities. The **lateral band** or medial calcaneocuboid ligament (12), less thick than the previous ligament, runs horizontally to be attached to the dorsal aspect of the cuboid. These two bands therefore form a solid right angle, which faces superiorly and laterally (fig. 15: diagrammatic sketch seen from in front).

The dorsal calcaneocuboid ligament (13) is a thin band (figs. 13 and 16) running from the supero-lateral aspect of the calcaneocuboid joint.

The plantar calcaneocuboid ligament which is dense and pearly white, runs over the plantar aspect of the tarsal bones. It consists of two distinct layers:

A deep layer (14) which unites (fig. 14: seen from below; the superficial layer has been cut and retracted) the anterior tubercle of the calcaneus to the plantar surface of the cuboid *just posterior to the groove for the peroneus longus tendon* (PL). This layer is often called the short plantar ligament.

A superficial layer ′(15), attached posteriorly to the plantar surface of the calcaneus between its posterior tubercles and its anterior tubercle and anteriorly to the plantar surface of the cuboid *anterior to the groove* for the peroneus longus (PL) tendon. It sends expansions to be inserted into the bases of the last four metatarsals. Thus the groove on the cuboid is transformed into a fibro-osseous tunnel which the peroneus longus tendon (PL) traverses lateromedially (17, fig. 16). Figure 18 (two paramedian sections viewed from inside) and figure 19 (showing the direction of these two sections) show the PL tendon leaving the cuboid groove. This band is also called the long plantar ligament.

This plantar calcaneocuboid ligament is one of the essential structures for the support of the arches of the foot (p. 202).

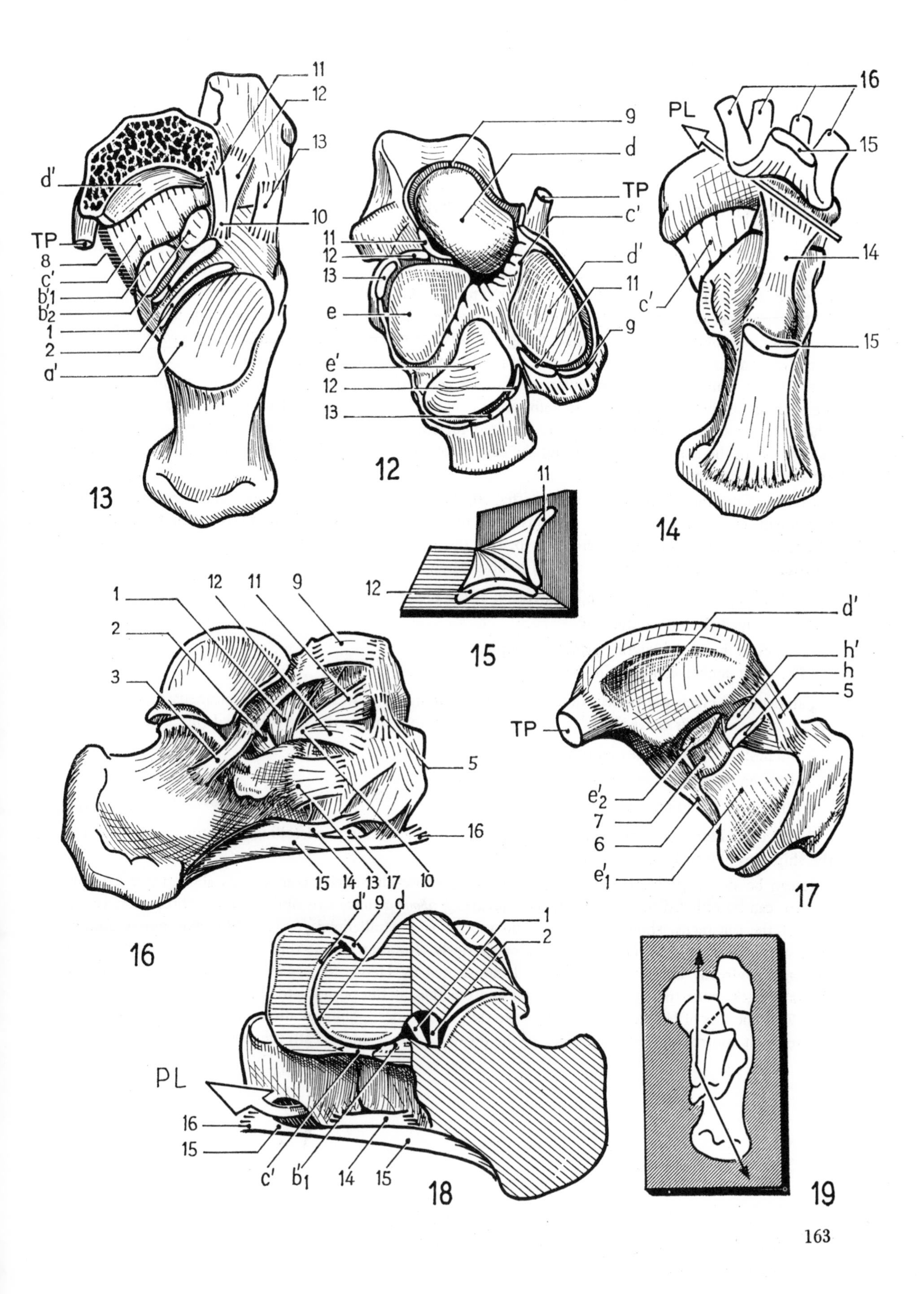

11
12
13
d'
10
TP
8
c'
b'1
b'2
1
2
a'
13
9
d
TP
c'
11
12
13
d'
11
e
9
e'
12
13
12
16
PL
15
14
c'
15
14
11
12
15
12
11
9
1
2
3
5
16
15
14
13
17
10
16
d'
h'
h
5
TP
e'2
7
6
e'1
17
d'
9
d
1
2
PL
16
15
c'
b1
14
15
18
19

THE MOVEMENTS OF THE SUBTALAR JOINT

Taken separately each of the surfaces of the subtalar joint can be roughly approximated to a geometrical surface: the superior surface of the calcaneus is a segment of a cylinder, the head of the talus is a segment of a sphere. However, this joint must be considered to be a **plane joint**, because it is geometrically impossible for two spherical and two cylindrical surfaces (contained within a single mechanical joint) to slide simultaneously on one another without contact being lost between one of the sets of surfaces concerned. The joint possesses some measure of 'play' by virtue of its structure and stands in sharp contrast to a very tight joint (e.g. the hip joint), where the articular surfaces are geometrically congruent and allow minimal 'play'. If the surfaces of the subtalar joint are sufficiently congruent in the intermediate position, i.e. the position where the greatest degree of contact is required for support of the body weight, they become frankly *incongruent* in the extreme positions and the area of contact is reduced, but then the stresses on the joint are also reduced appreciably.

Starting from the intermediate position (fig. 20: the 'transparent' calcaneus and talus seen from inside), movements of the calcaneus on the talus (assumed to be fixed) occur **simultaneously in the three planes of space**. During **inversion of the foot** (p. 156) the anterior extremity of the calcaneus undergoes **three elementary movements** (fig. 21: initial position shown in dashes):

it moves slightly distally (t): slight extension of the foot;

it moves medially (v): adduction;

it comes to lie on the lateral surface (r): supination.

(The exact converse applies in the movement of eversion.)

Farabeuf has given the perfect description of this complex movement of the calcaneus: '*the calcaneus pitches, turns and rolls under the talus*'. This comparison to a ship is perfectly justified (fig. 24):

it pitches: its stem plunges into the waves (a);

it turns on itself (b);

it rolls by tilting to one side (c).

These elementary movements about the axes of pitching, turning and rolling are automatically fused as the ship dips obliquely into the wave (e).

It can be shown geometrically that a movement, whose elementary components about three axes are known, can be reduced to a *single movement occurring about a single axis* oblique to the three axes. In the case of the calcaneus, shown here diagrammatically as a paralleliped (fig. 22), this axis mn is oblique supero-inferiorly, mediolaterally and anteroposteriorly. Rotation about this axis (fig. 23) results in the movements already described. This axis, demonstrated by Henke, enters at the superomedial aspect of the talar neck, runs through the sinus tarsi and emerges at the posterolateral tubercle of the calcaneus (p. 170). The axis of Henke not only applies to the talocalcanean joint but also to the transverse tarsal joint. Therefore it controls all the movements of the posterior tarsus underneath the ankle.

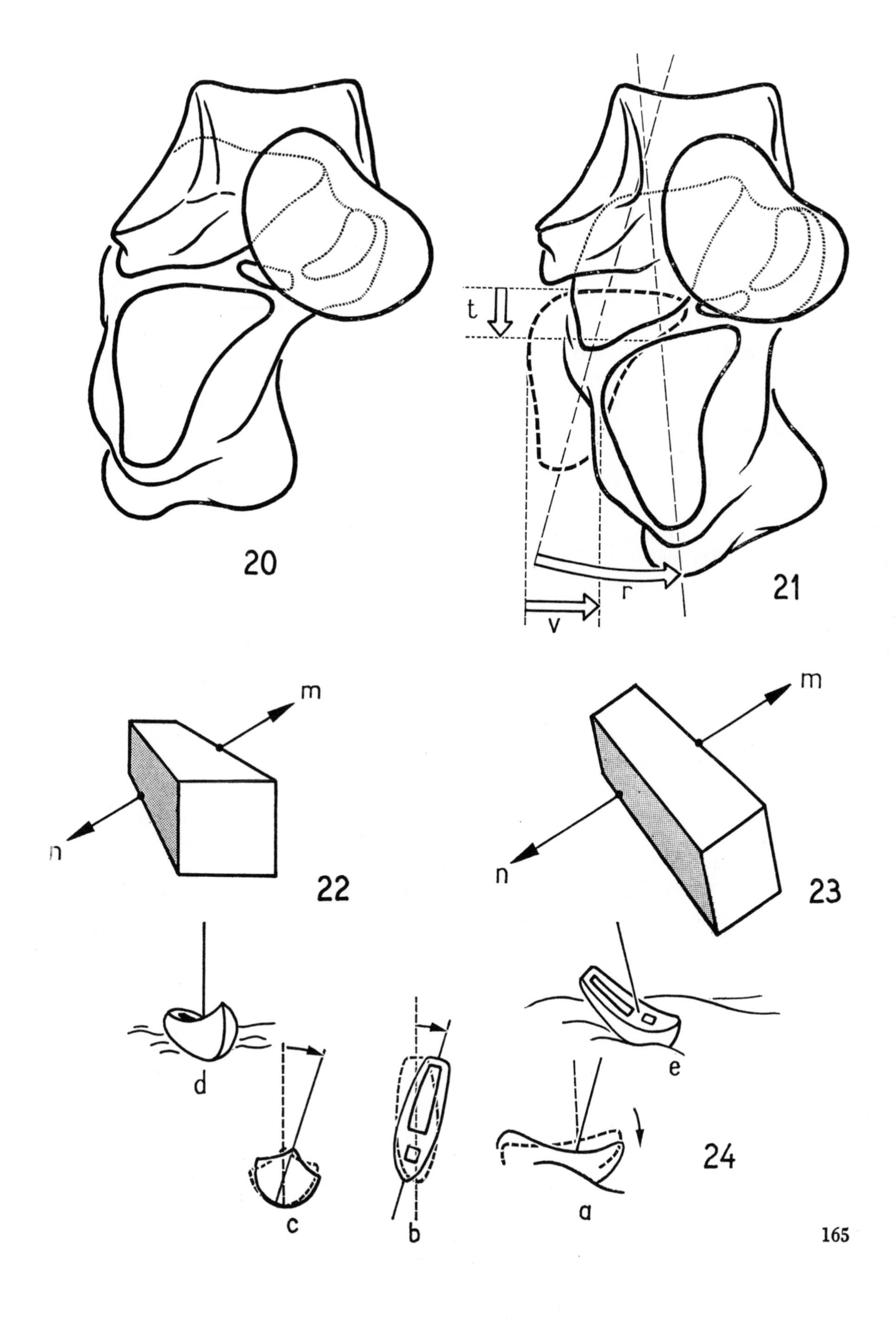
20
t
v
r
21
m
n
22
m
n
23
d
c
b
e
a
24

THE MOVEMENTS OF THE SUBTALAR AND TRANSVERSE TARSAL JOINTS

The relative displacements of the bones of the posterior tarsus are easily analysed with the use of an anatomical preparation and skiagrams taken in the positions of inversion and eversion. If each bone is transfixed with a metal pin (a: for the talus; b: for the calcaneus; c: for the navicular; d: for the cuboid) the angular displacements can be calculated.

On a **skiagram taken vertically** (seen from above), the calcaneus staying put, the change from eversion (fig. 25) to inversion (fig. 26) is associated with the following displacements:

the navicular (c) slides medially on the talar head and turns through an angle of 5°;

the cuboid (a) follows the navicular and turns through the same angle and slides medially relative to the calcaneus and the navicular;

the calcaneus (b) moves anteriorly slightly and turns on the talus through an angle of 5°.

These three elementary rotations occur in the same direction, i.e. in *adduction.*

An **anteroposterior skiagram,** the talus being taken to be stationary, shows the following displacement during change from eversion (fig. 27) to inversion (fig. 28):

the navicular (c) turns through an angle of 25° and 'overflows' the talus medially;

the cuboid (d) is completely lost behind the shadow of the calcaneus and turns through an angle of 18°;

the calcaneus (b) slides medially under the talus and turns through an angle of 20°.

These three elementary rotations occur in the same direction i.e. that of *supination,* and the navicular turns more than the calcaneus and especially more than the cuboid.

Finally, on **a lateral view,** during change from eversion (fig. 29) to inversion (fig. 30) the following displacements are noted:

The navicular (b) literally slides under the talar head and turns on itself through an angle of 45° so that its anterior surface tends to face inferiorly.

The cuboid (d) also slides inferiorly in relation to both the calcaneus and the talus; this inferior movement of the cuboid with respect to the talus is distinctly more important than that of the navicular relative to the talus. At the same time the cuboid turns through an angle of 12°.

The calcaneus (b) finally moves anteriorly relative to the talus so that the posterior edge of the talus comes to overhang the part of the calcaneus lying posterior to its superior articular facet. At the same time it turns through an angle of 12° in the direction of extension, like the navicular.

These three elementary movements occur in the same direction, i.e. that of extension.

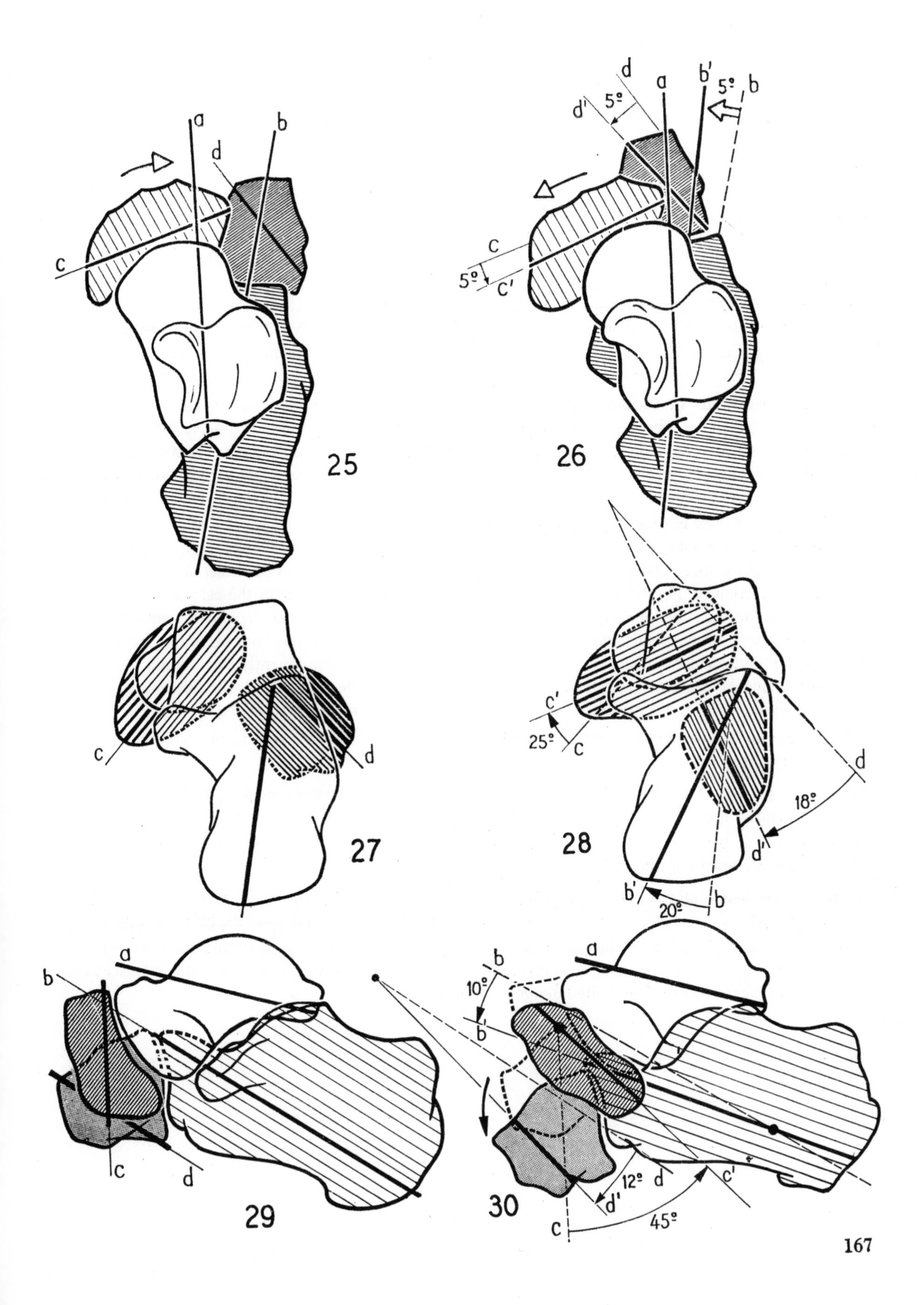
a
b
d
c
25
d
a
b'
5º
b
d'
5º
c
5º
c'
26
c
d
27
c'
25º
c
d
18º
28
d'
b'
20º
b
a
b
c
d
29
a
b
10º
b'
12º
d
c'
d'
30
c
45º

THE MOVEMENTS AT THE TRANSVERSE TARSAL JOINT

These movements depend on the shape of the articular surfaces and the arrangement of the ligaments.

Taken as a whole, (fig. 31) the articular surfaces are set in relation to an axis xx′ which runs obliquely supero-inferiorly and lateromedially at an angle of 45° with the horizontal. It serves as a '*hinge*' which allows the navicular and the cuboid to move inferiorly and medially (arrows S and C) or superiorly and laterally. The surface of the talar head, which is oval with its great axis yy′ inclined at an angle of 45° with the horizontal (the angle of 'rotation' of the talus), is longer in the direction of this movement.

The displacements of the navicular on the head of the talus take place medially (fig. 32) and inferiorly (fig. 33) under the pull of the tibialis posterior (TP) which is inserted into the tubercle of the navicular. The tension of the dorsal talonavicular ligament (a) checks these movements. This change in the direction of the navicular produces, via the cuneiform bones and the first three metatarsals, adduction and hollowing of the medial plantar arch (p. 200).

At the same time **the navicular moves relative to the calcaneus.** In the *position of eversion* (fig. 34: seen from above; the talus has been removed) the 'spring' ligament (b), the lower edge of the deltoid ligament (c) and the medial band of the bifurcated ligament (d) are under tension; *when the foot is inverted* (fig. 35), contraction of the tibialis posterior (TP) brings the navicular closer to the calcaneus and makes the talus move up the superior surface of the calcaneus (striped arrow) so that the above-mentioned ligaments are relaxed. This explains why the anterior surface of the calcaneus does not extend as far down as the navicular: an articular surface, supported by a bony bracket, and consequently rigid, would not allow these displacements of the navicular relative to the calcaneus. On the other hand, the flexible surface of the spring ligament (b) is essential (p. 200) for the elasticity of the medial arch of the foot.

The superior movements of the cuboid on the calcaneus (fig. 36: seen from inside) is severely limited by two factors:

the anterior process of the calcaneus (arrow) which constitutes an obstacle on the proximal side of the transverse tarsal joint;

the tension of the powerful plantar calcaneocuboid ligament (f) which rapidly stops the interspace of the joint (α) from opening out inferiorly.

On the other hand, inferior movement of the cuboid (fig. 37) easily takes place over the convexity of the articular facet of the calcaneus; it is only checked by tension of the lateral band (e) of the bifurcated ligament.

In the transverse plane (fig. 38: horizontal section at the level AB of fig. 31) the cuboid glides more easily medially being checked by tension of the dorsal calcaneocuboid ligament (g). Taken as a whole, displacement of the cuboid takes place preferentially *inferiorly and medially.*

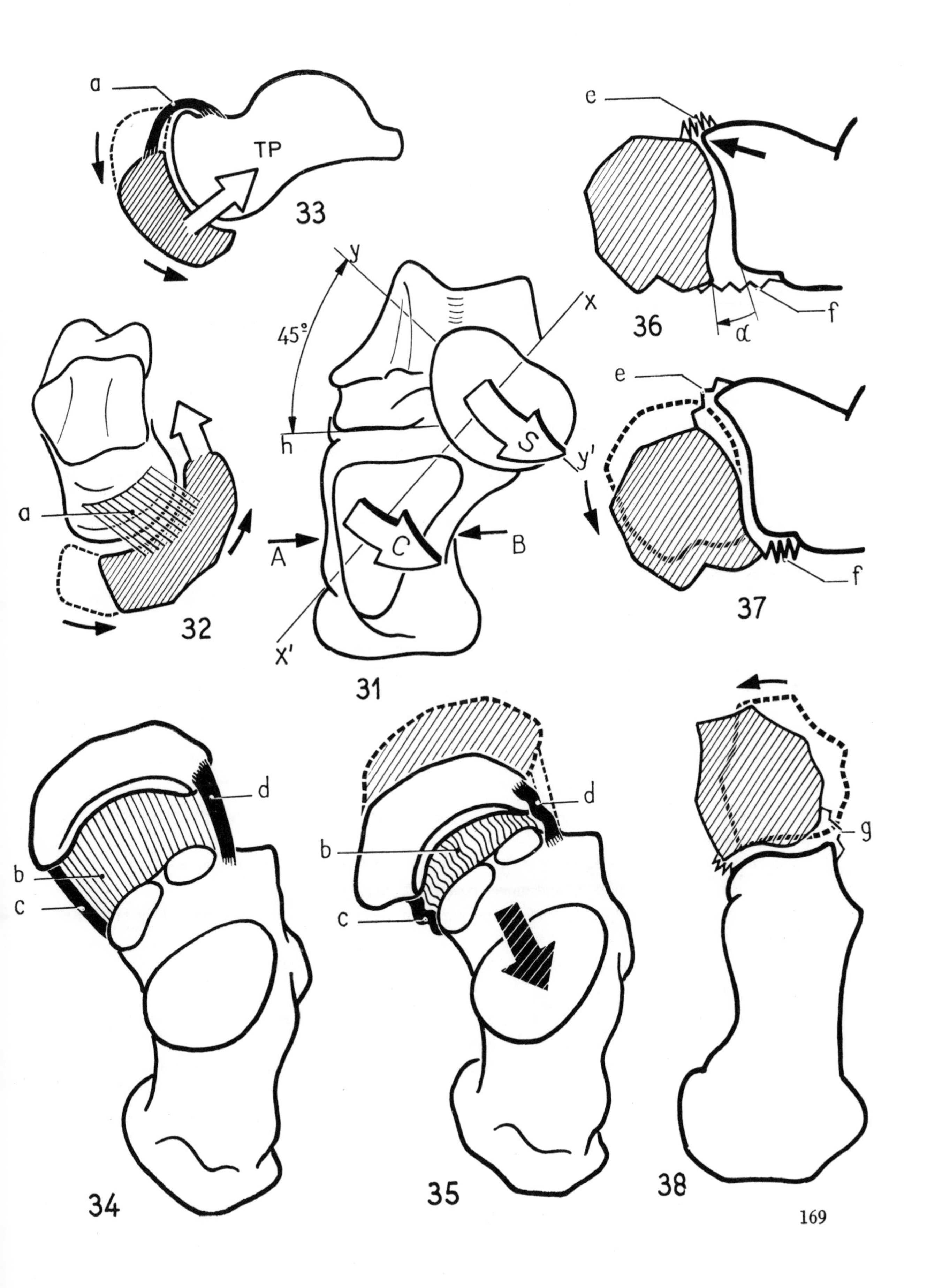

a
TP
33
e
36
f
α
y
x
45°
h
S
y'
a
A
C
B
32
x'
31
e
37
f
d
b
c
34
d
b
c
35
g
38

THE OVERALL FUNCTION OF THE JOINTS OF THE POSTERIOR TARSUS

(the legends are the same as those of p. 162).

It is clear from examining and manipulating an anatomical preparation of the posterior part of the foot that all these joints constitute a *single functional unit*, concerned with altering the direction and shape of the arches of the foot. The subtalar and the transverse tarsal joints are mechanically linked and together form a single joint with **one degree of freedom** about the axis of Henke (mn).

The diagrams show the four bones of the posterior part of the foot from two different angles: from the front and from the outside (figs. 39 and 41); from the front (figs. 40 and 42). In these diagrams the positions corresponding to inversion (figs. 39 and 40) and those corresponding to eversion (figs. 41 and 42) are correspondingly arranged one above the other. This makes it easier to appreciate the changes in direction of the navicular and cuboid bones relative to the talus, which by definition stays put.

Movement of inversion (figs. 39 and 40):

the tibialis posterior pulls on the navicular (nav) exposing the superolateral part of the talar head (d);

the navicular drags the cuboid along via the cubonavicular ligaments;

the cuboid in turn pulls on the calcaneus (Calc) which moves anteriorly under the talus (Tal);

the sinus tarsi opens to its widest (fig. 39) while the two bands of the interosseous ligament (1 and 2) are stretched;

the superior articular facet of the calcaneus (a′) is laid bare on its antero-inferior aspect while the interspace of the subtalar joint gapes open superoposteriorly.

Taken as a whole:

the navicular and cuboid bones together are drawn medially (fig. 40: arrow Add) so that the forefoot *moves anteriorly and medially* (arrow I, fig. 39).

At the same time *the pair scaphoid-cuboid turns round an anteroposterior axis running through the bifurcated ligament*, which actively resists stresses of torsion and traction. This rotation, due to the superior displacement of the scaphoid and the inferior movement of the cuboid, produces **supination** (arrow Supin): the sole of the foot faces medially because the lateral plantar arch is lowered—the articular facet of the cuboid, corresponding to the base of the fifth metatarsal (Vm) faces anteriorly and inferiorly—while the medial arch is elevated—the articular facet of the navicular for the medial cuneiform (Ic) moves anteriorly.

Movement of eversion (figs. 41 and 42):

the peroneus brevis, inserted into the tuberosity of the base of the fifth metatarsal, pulls the cuboid laterally and posteriorly;

the cuboid draws the navicular along so that the superomedial part of the talar head is exposed;

the calcaneus is also drawn along and moves posteriorly under the talus;

the sinus tarsi closes down (fig. 41) and the movement of eversion is checked by impact of the talus on to the floor of the sinus tarsi;

the posterosuperior part of the superior surface of the calcaneus (a′) is uncovered.

Taken as a whole:

the navicular and cuboid bones (fig. 42) together are pulled laterally (arrow Abd) so that the forefoot is drawn *anteriorly and laterally* (arrow E, fig. 41).

At the same time it rotates in the direction of **pronation** (arrow Pron) as a result of inferior displacement of the navicular and abduction of the cuboid so that its articular facet for the fifth metatarsal (Vm) looks anteriorly and laterally.

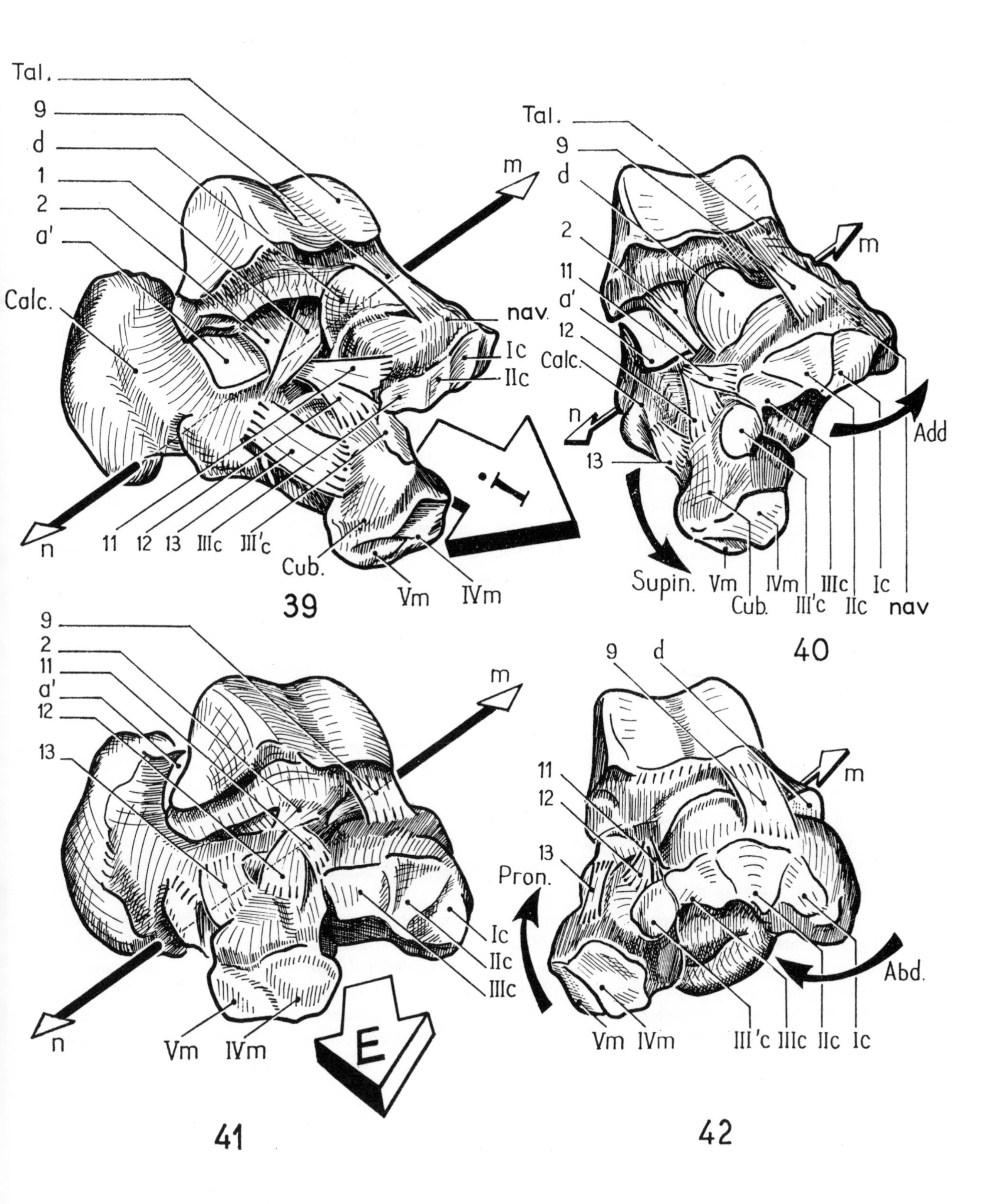
Tal.
9
d
1
2
a'
Calc.
m
nav.
Ic
IIc
n
11
12
13
IIIc
III'c
Cub.
Vm
IVm
I
39
Tal.
9
d
2
11
a'
12
Calc.
m
n
13
Add
Supin.
Vm
Cub.
IVm
III'c
IIIc
IIc
Ic
nav
40
9
2
11
a'
12
13
m
n
Ic
IIc
IIIc
Vm
IVm
E
41
9
d
11
12
13
Pron.
m
Abd.
Vm
IVm
III'c
IIIc
IIc
Ic
42

THE CUNEONAVICULAR, INTERCUNEIFORM AND TARSOMETATARSAL JOINTS

(the numbers and letters have the same meanings as on pp. 162 and 170).

All these joints are plane joints; their articular surfaces glide on each other and move apart so that the interspace of the joint gapes slightly.

The navicular bone (fig. 43: seen from in front) shows *three articular facets* (I_c, II_c, III_c), which articulate with the medial, intermediate and lateral cuneiforms respectively. The cuboid also shows three facets which articulate with the fifth metatarsal (Vm), the fourth metatarsal (IVm), and the lateral cuneiform (III′C); the cuboid also supports the lateral edge of the navicular (the cubonavicular joint: white arrows).

Figure 44 (an anterolateral view in perspective) illustrates how the three cuneiforms (C_m, C_i, C_l) articulate with the navicular and the cuboid: the double arrow indicates how the lateral cuneiform rests on the cuboid, i.e. on a facet (III′C) lying just in front of the articular facet for the navicular (cuneocuboid joint).

The intercuneiform joints (fig. 45: the cuneonavicular, the intercuneiform and some tarsometatarsal joints are viewed from above) comprise articular facets and interosseous ligaments: between the medial and intermediate cuneiforms the interosseous ligament (19) has been cut; between the intermediate and lateral cuneiforms the ligament (20) is still intact.

The tarsometatarsal joints comprise, on the proximal side (fig. 47: seen from above), the three cuneiforms (C_m, C_i, C_l) medially and the cuboid (Cub) laterally and, on the distal side, the bases of the five metatarsals (M_I, M_{II}, M_{III}, M_{IV}, M_V). They are all plane joints which overlap one another very closely. A dorsal view of the opened joints (fig. 46, according to Rouvière) shows the various articular facets of the tarsal bones and the corresponding surfaces of the metatarsals. The base of the second metatarsal (M_{II}) fits into the cuneiform mortise formed by the lateral facet (IIm C_m) of the medial cuneiform (C_m), the anterior facet (IIm C_i) of the intermediate cuneiform (C_i) and the internal facet (II_m C_l) of the lateral cuneiform (C_l). The tarsometatarsal joints are supported by powerful ligaments which become visible (fig. 45), when the joints are opened from above, the first metatarsal is rotated on its axis (arrow I) and the third is pulled out laterally (arrow 2). These ligaments are:

Medially, the **powerful dorsal ligament** running from the lateral aspect of the medial cuneiform to the medial surface of the base of the second metatarsal (M_{II}). It is the key ligament during surgical disarticulation.

Laterally, a series of **dorsal ligaments** including straight fibres (21) between the intermediate cuneiform (C_i) and the second metatarsal (M_{II}) and (22) between the lateral cuneiform (C_l) and the third metatarsal (M_{III}) and cruciate fibres (23) between the lateral cuneiform and the second metatarsal and (24) between the intermediate cuneiform and the third metatarsal.

The strength of the tarsometatarsal joints also depends on numerous ligaments (fig. 47: dorsal view; fig. 48: plantar view) running from the base of each metatarsal to the corresponding tarsal bone and between the bases of the adjacent metatarsals. These include, in particular on the dorsal aspect (fig. 47), the ligaments radiating from the base of the second metatarsal to all the neighbouring bones and, on the plantar aspect (fig. 48), the ligaments between the medial cuneiform and the first three metatarsals. To the plantar aspect of the base of the first metatarsal is attached the tendon of the peroneus longus (PL) as it emerges from its plantar groove (broken line 25).

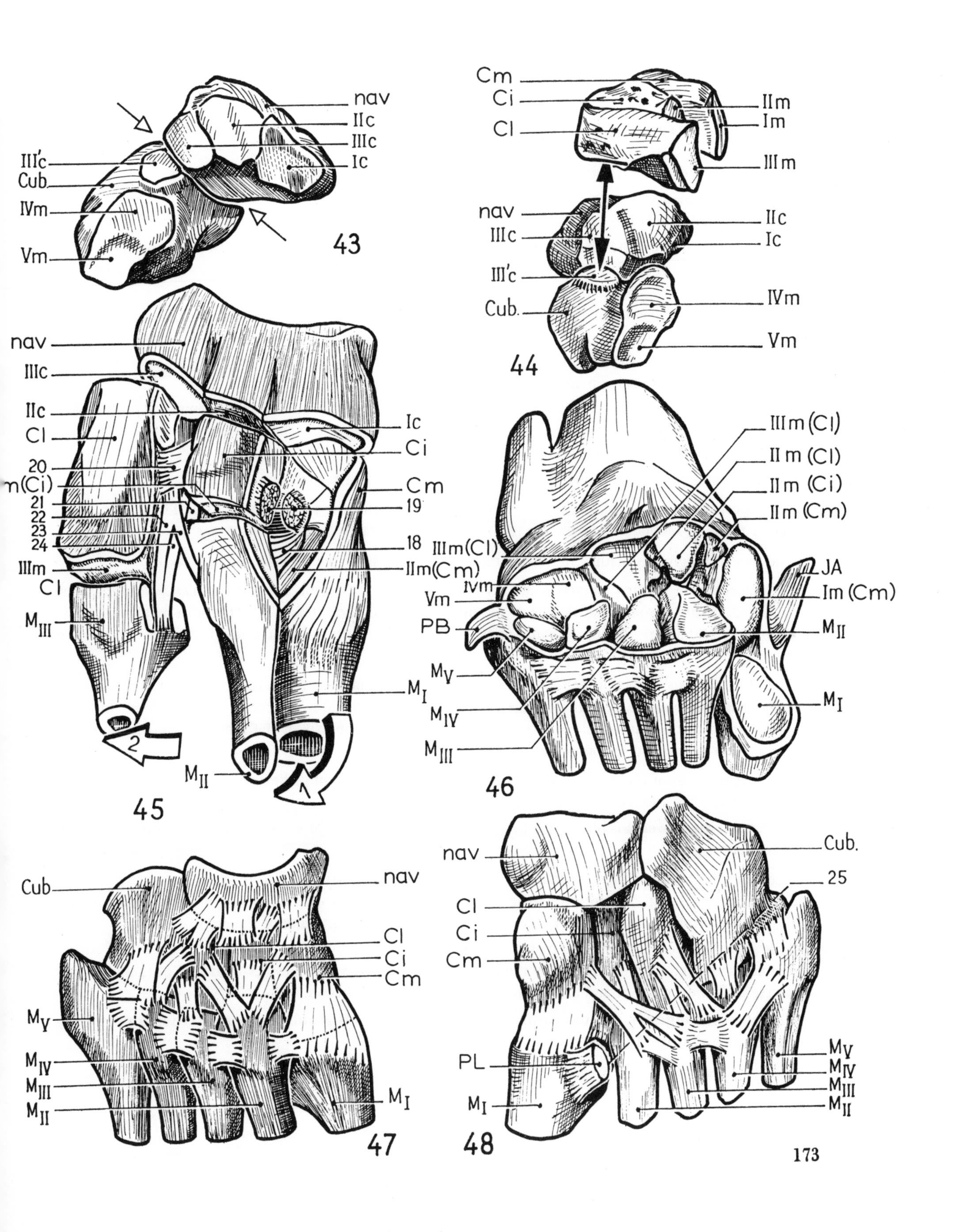
nav
IIc
IIIc
Ic
III'c
Cub
IVm
Vm
43
Cm
Ci
Cl
IIm
Im
IIIm
nav
IIIc
IIc
Ic
III'c
Cub.
IVm
Vm
44
nav
IIIc
IIc
Cl
20
m(Ci)
21
22
23
24
IIIm
Cl
Ic
Ci
Cm
19
18
IIm(Cm)
2
45
IIIm (Cl)
IIm (Cl)
IIm (Ci)
IIm (Cm)
IIIm(Cl)
IVm
Vm
PB
JA
Im (Cm)
46
Cub
nav
Cl
Ci
Cm
47
nav
Cub.
25
Cl
Ci
Cm
PL
48

MOVEMENTS OF THE ANTERIOR TARSAL AND TARSOMETATARSAL JOINTS

The **intercuneiform joints** (fig. 49: frontal section) allow small vertical movements to occur which alter the curvature of the transverse plantar arch (p. 206). The lateral cuneiform (C_l) rests on the cuboid (Cub) whose medial third (striped) provides support for the cuneiform arch.

Slight displacements of the cuneiforms relative to the navicular (nav) occur along the long axis of the foot (fig. 50: sagittal section) and contribute to the changes of curvature of the medial arch (p. 200).

The movements of the tarsometatarsal joints can be deduced from their anatomical features, especially the shape of the joint interspaces and the orientation of the articular surfaces (fig. 51: seen from above):

As a whole the line of the tarsometatarsal joints runs *obliquely mediolaterally*, *supero-inferiorly* and *anteroposteriorly*. Its medial end lies 2 cm. anterior to its lateral end. The general obliquity of this axis of flexion and extension of the tarsometatarsal joints contributes, just as the obliquity of the axis of Henke, to the movements of eversion and inversion.

The distances by which the cuneiforms overstep one another and the cuboid are in geometric progression:

the lateral cuneiform (C_l) oversteps the cuboid (Cub) by 2 mm.;

the lateral cuneiform (C_l) oversteps the intermediate cuneiform (C_i) by 4 mm;

the medial cuneiform (Cm) oversteps the intermediate cuneiform by 8 mm.

Thus is constituted the deep mortise which lodges the base of the second metatarsal. This metatarsal is therefore the least mobile and forms the crest-tile of the plantar arches (p. 204).

The two end segments of this line of the metatarsal joints have the opposite obliquity:

The interspace of the first metatarso-medial cuneiform joint is oblique anteriorly and *laterally* and, when produced, it runs through the middle of the fifth metatarsal; the interspace of the fifth metatarsocuboid joint is oblique anteriorly and *medially* and, when produced, runs almost through the head of the first metatarsal.

Therefore the axis of flexion and extension of the lateral metatarsals, which are the most mobile, is not perpendicular but oblique to their long axes. So these metatarsals do not move in a sagittal plane but over the segment of a cone: during flexion they also move towards the axis of the foot (fig. 53: diagrammatic representation of two end tarsometatarsal joints):

the movement aa′ of the head of M_I is compounded of a movement of flexion (F) and one of abduction (Abd), which has a range of 15° (Fick);

conversely, the movement bb′ of the head of M_V is compounded of a movement of flexion (F) and one of adduction (Add).

Therefore the heads of these metatarsals not only move inferiorly but also towards the axis of the foot and this **increases** (fig. 54) **the curvature of the anterior arch** with hollowing of the anterior part of the foot. Conversely, extension of the metatarsals is followed by flattening of the arch.

This approximation of the lateral metatarsals is also assisted (fig. 52: the articular surfaces of the cuboid and cuneiforms seen from the front) by the obliquity of the transverse axes (xx′ and yy′) of their articular surfaces. This movement occurs along the thick double-headed arrows.

Therefore changes in the curvature of the anterior arch result from movements occurring at the tarsometatarsal joints

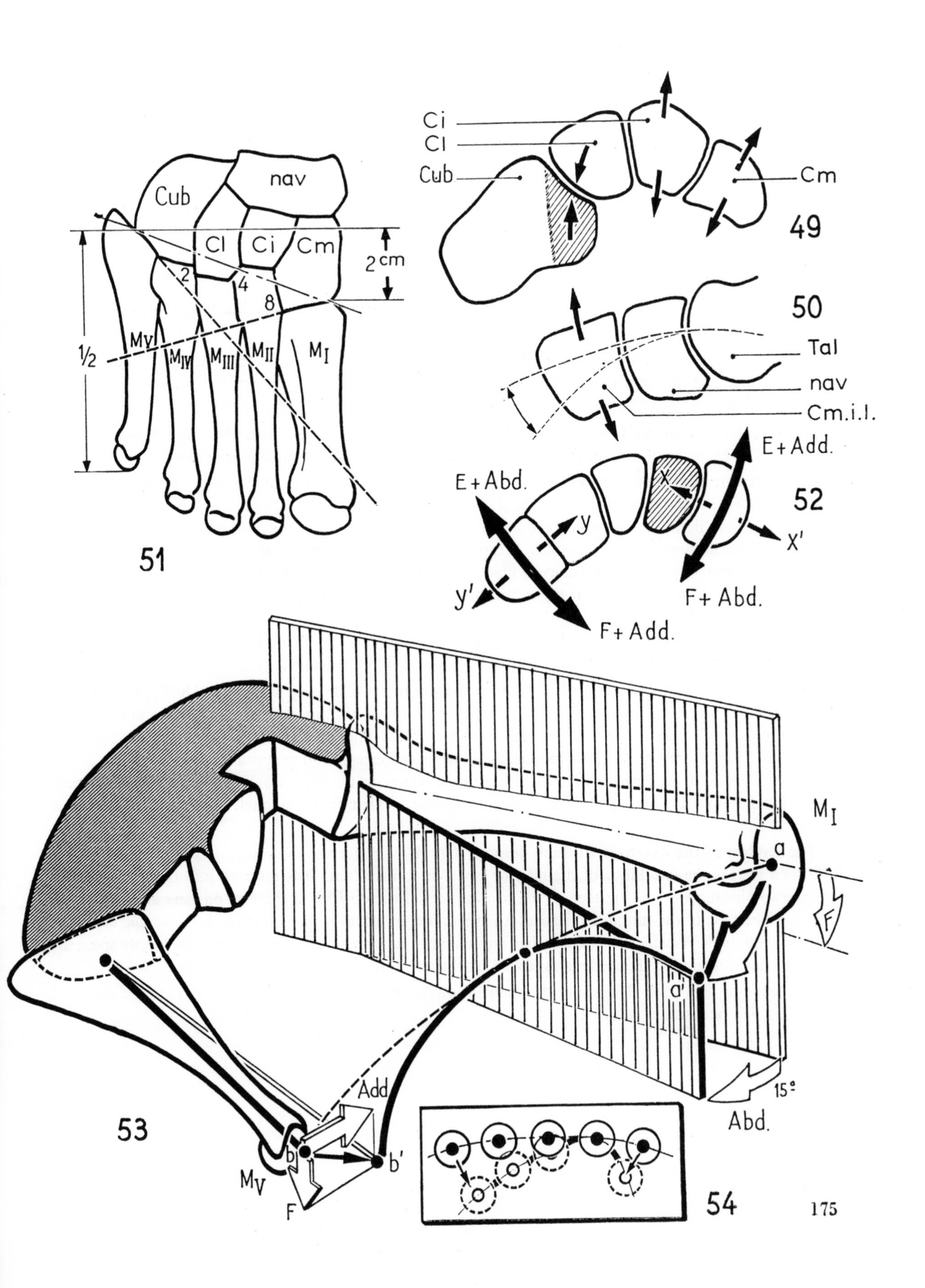

Ci
Cl
Cub
Cm
49
50
Tal
nav
Cm.i.l.
E+Add.
E+Abd.
52
y
y'
x'
F+Abd.
F+Add.
Cub
nav
Cl
Ci
Cm
2 cm
2
4
8
½
M_V
M_{IV}
M_{III}
M_{II}
M_I
51
M_I
a
F
a'
15º
Abd.
Add
53
b
b'
M_V
F
54

EXTENSION OF THE TOES

The metatarsophalangeal and interphalangeal joints will not be described as they are identical to those of the fingers (see Vol. I), except for some functional differences. Thus, whereas at the metacarpophalangeal joints flexion has a greater range than extension, extension **exceeds** flexion at the metatarsophalangeal joints:

active extension has a range of 50° to 60°, active flexion only 30° to 40°;

passive extension, which is essential in the final phase of taking a step (fig. 55), reaches or exceeds 90° while passive flexion has a range of 45° to 50°.

Side-to-side movements of the toes occur at the metatarsophalangeal joints and have a far smaller range than those of the fingers. In particular man's big toe, in contrast to that of the monkey, has lost all *movement of opposition* during adaptation to walking on two legs.

Active extension of the toes is produced by *three muscles*: two extrinsic muscles—the extensor hallucis longus and extensor digitorum longus—and one intrinsic muscle—extensor digitorum brevis.

The **extensor digitorum brevis** (fig. 56) lies entirely in the dorsum of the foot. It arises from the sulcus calcanei (i.e. the floor of the sinus tarsi) and from the stem of the inferior extensor retinaculum. It divides into four fleshy bellies which are inserted by tendon into the lateral sides of the corresponding tendons of the extensor digitorum longus, except for the tendon (destined for the first metatarsal), which is inserted directly into the dorsal surface of first phalanx of the big toe; the fifth toe receives no tendon from this muscle. Therefore the extensor digitorum brevis is an extensor of the metatarsophalangeal joints of the first four toes (fig. 57).

The extensor digitorum longus and the extensor hallucis longus are lodged in the anterior compartment of the leg; their tendons terminate on the phalanges (see p. 178 for details).

The extensor digitorum longus (fig. 58) runs anterior to the ankle deep to the lateral half of the superior extensor retinaculum and then passes posterior to the stem of the inferior extensor retinaculum before dividing into four tendons which run to the four lesser toes. Therefore the fifth toe is only extended by the extensor digitorum longus. This muscle, as its name indicates, is an extensor of the toes but it is *also and above all* (p. 184) *a flexor of the ankle.* Its primary action on the toes is only apparent when its flexor action on the ankle is counterbalanced by the antagonistic extensor of the ankle (the Achilles tendon is shown as an arrow).

The **tendon of the extensor hallucis longus** (fig. 59) runs deep to the superior extensor retinaculum and then pierces the two limbs of the inferior extensor retinaculum. It is inserted into the dorsal aspect of the two phalanges of the big toe: on the two borders of the dorsum of the first phalanx and the dorsal aspect of the base of the terminal phalanx. It is therefore an extensor of the big toe but *also and above all a flexor of the ankle.* As with the extensor digitorum longus, its primary action on the big toe is only apparent when its flexor action on the ankle is cancelled by the antagonistic extensors of the ankle.

Duchenne de Boulogne claims that the extensor digitorum brevis is the only true extensor of the toes; this claim will be justified later.

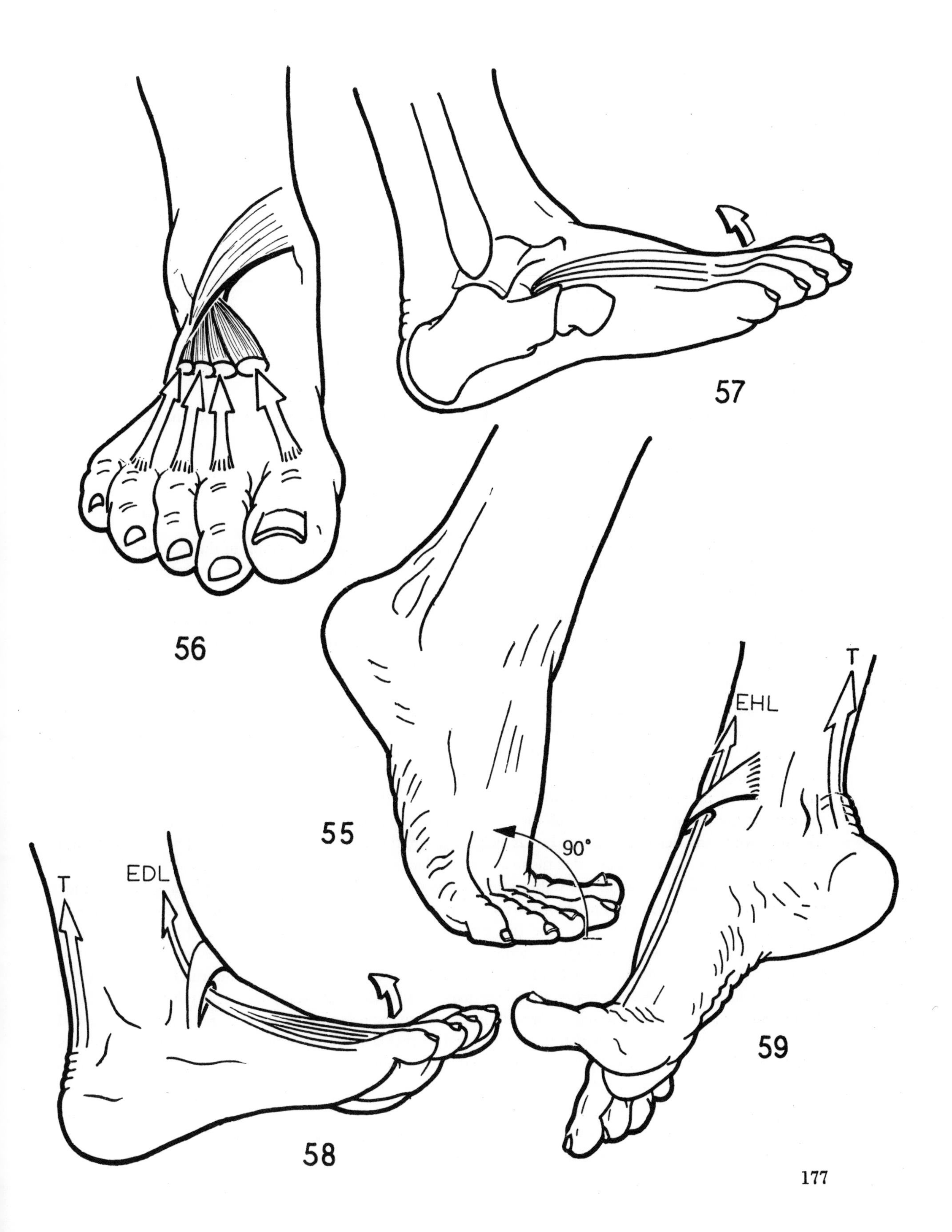
56
57
55
90°
T
EDL
T
EHL
58
59

THE INTEROSSEOUS AND THE LUMBRICAL MUSCLES

(the numbers have the same meaning in all the diagrams).

As in the hand, the interossei fall into two groups—dorsal and plantar—but their arrangement is slightly different in the foot (fig. 60: frontal section of the foot, posterior slice shown). The **four dorsal interossei** (Ix. d) are, as it were, centred on the second metatarsal (instead of the third as in the hand) and are inserted into the second toe and the two immediately adjacent toes (fig. 67: white arrows). The **three plantar interossei** (Ix. p) arise from the medial aspect of the plantar surfaces of the last three metatarsals and are inserted into the corresponding toe (fig. 68).

The mode of insertion of the interossei of the foot is similar to that of the interossei of the hand (fig. 61: dorsal view of the extensor tendons; fig. 63: side view of the muscles of the toes):

they are inserted into the medial or lateral side of the first phalanges (1);

they are also attached to the collateral digital expansions (3) of the extensor tendons by a tendinous expansion (2).

The tendon of the extensor digitorum longus (EDL) is inserted into the toes just as the extensor digitorum communis is into the fingers:

by some fibres (4) into the borders of the first phalanx (P_1) and not into its base;

by a median dorsal expansion (5) into the base of P_2;

by two collateral expansions (3) into the base of P_3.

Proximal to the first metatarsophalangeal joint (fig. 62: dorsal view) the tendons of the extensor digitorum longus for the second, third and fourth toes receive the corresponding tendons from the extensor digitorum brevis (EDB).

As in the hand, there are **four lumbricals** (figs. 60, 62 and 72) arising from the tendons of the flexor digitorum longus (the homologue of flexor digitorum profundus of the hand). Each lumbrical runs *medially* (fig. 72) to be inserted (figs. 62 and 63) like an interosseus, i.e. into the base of P_1 (6) and into the collateral expansion of the extensor longus (7).

The **tendon of the flexor digitorum longus** (FDL), like the flexor digitorum profundus in the hand (figs. 63 and 72), runs against the fibrocartilaginous plate (8) of the first metatarsophalangeal joint, and then 'perforates' the tendon of the flexor digitorum brevis (FDB) before gaining insertion into the base of the distal phalanx. The flexor digitorum brevis is therefore analogous to the flexor digitorum sublimis: it is superficial and its tendon is '*perforated*' by the tendon of the long flexor prior to its insertion into the margins of the second phalanx. Thus the flexor digitorum longus flexes the distal interphalangeal joint (fig. 65), while the flexor digitorum brevis flexes the proximal interphalangeal joint. The interossei and lumbrical muscles, as in the hand, flex the metatarsophalangeal joint (fig. 64) and extend the interphalangeal joints. They play a vital part in the stabilisation of the toes: by flexing the metatarsophalangeal joint they provide a strong point of attachment for the extensors of the toes as they flex the ankle. When the interossei and lumbricals are paralysed a 'claw foot' (fig. 66) can result: as the metatarsophalangeal joint is no longer stabilised by the interossei, it is hyperextended by the extensors and the phalanx slides over the dorsal surface of the head of the metatarsal. The foot is then secondarily fixed in this abnormal position by the dorsal displacement of the interossei above the axis (+) of the metatarsophalangeal joint. Furthermore the interphalangeal joints are flexed as a result of the 'relative shortening' of the flexors and this is followed by dorsal subluxation of the proximal interphalangeal joint (black arrow) between the collateral expansions of the extensor tendon so that the action of the extensor is now reversed.

As in the hand, the position of the toes depends therefore on the balance struck among different muscles. Thus it becomes apparent, as claims Duchenne de Boulogne, that only the extensor digitorum brevis is the true extensor of the toes; the extensors are in actual fact flexors of the ankle and so, according to Duchenne, would be more 'efficient' if they were inserted directly into the metatarsals.

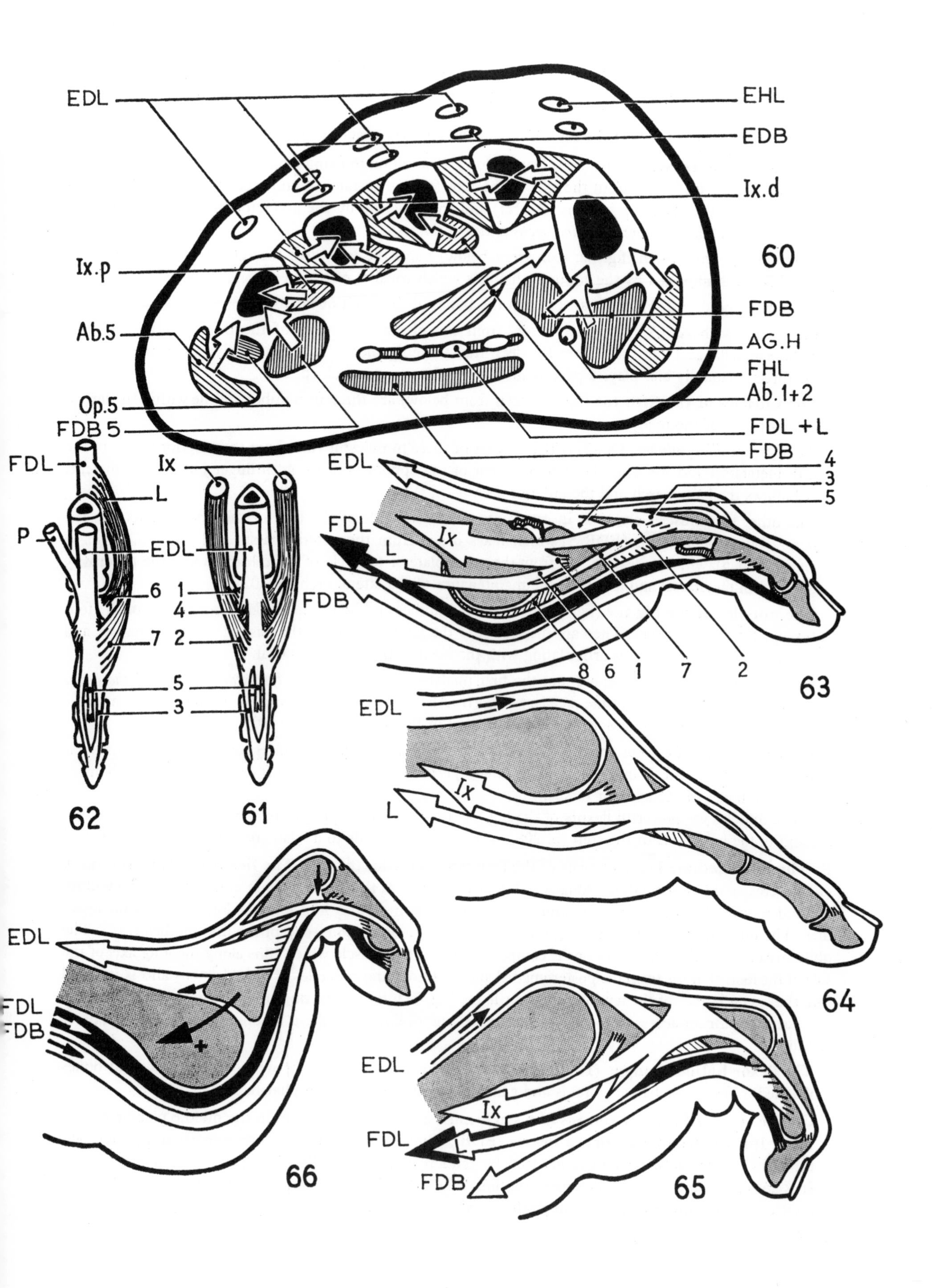

EDL
EHL
EDB
Ix.d
Ix.p
60
FDB
Ab.5
AG.H
FHL
Ab.1+2
Op.5
FDB 5
FDL + L
FDB
FDL
L
P
EDL
6
7
5
3
Ix
1
4
2
62
61
EDL
FDL
Ix
L
FDB
4
3
5
8
6
1
7
2
63
EDL
Ix
L
64
EDL
FDL
FDB
66
EDL
Ix
FDL
L
FDB
65

THE SOLE OF THE FOOT: THE PLANTAR MUSCLES

(the numbers and letters have the same meaning as in the diagrams of the previous page).

A. The **deep layer** consists of the interossei and the muscles attached to the fifth toe and to the big toe:

The dorsal interossei (fig. 67: seen from below), in addition to being flexors and extensors of the toes, also abduct the toes away from the axis of the foot (second metatarsal and second toe). The big toe is 'abducted' by the abductor hallucis (Ab. H) and the little toe by the abductor digiti minimi (Ab. 5). These two muscles are therefore *analogous to dorsal interossei.*

The plantar interossei (fig. 68: seen from below) move the last three toes closer to the second toe. The big toe is adducted by the adductor hallucis which consists of two heads:

the oblique head (Ad. 1) which arises from the bones of the anterior tarsus;

the transverse head (Ad. 2) which arises from the plantar ligaments of the third, fourth and fifth metatarsophalangeal joints and from the deep transverse ligaments. It draws the first phalanx of the big toe directly laterally and plays a part in supporting the anterior arch (p. 204).

The muscles of the fifth toe (fig. 69: seen from below) are three in number and lie within the lateral compartment of the foot:

the *opponens digiti minimi* (Op. 5) is the deepest of these muscles: it runs from the anterior tarsus to the fifth metatarsal and has a similar action to that of the opponens of the fifth finger but is less efficient. It hollows the lateral arch and the anterior arch.

The other two muscles are both inserted into the lateral side of the base of the first phalanx. i.e. the *flexor digiti minimi brevis* (FDB 5) which takes origin from the anterior tarsus and the *abductor digiti minimi* (Ab. 5), which arises (fig. 70) from the posterolateral tubercle of the calcaneus and the tuberosity of the fifth metatarsal and assists in the maintenance of the lateral arch (p. 202).

The muscles of the big toe (fig. 69) are three in number and lie in the medial compartment of the foot (except for the adductor). They are inserted into the lateral aspect of the base of the first phalanx and into the two sesamoid bones articulating with the head of the first metatarsal.

On the medial side a sesamoid and the first phalanx give insertion to *the medial portion of the flexor hallucis brevis* (FHB) and to the *abductor hallucis* (Ab. H) which arises from the posteromedial tubercle of the calcaneus (fig. 70) and assists in the support of the medial arch (p. 202).

On the lateral side, a sesamoid bone and the first phalanx receive the insertion of the *two heads of the adductor hallucis* (Ad. 1 and 2) and of *the lateral portion of the flexor hallucis brevis* (FHB), which arises from the bones of the anterior tarsus.

These muscles are powerful flexors of the big toe: they play an important part in stabilisation of the big toe and in the last phase of the step (p. 208). Their paralysis leads to a 'claw' deformity of the big toe.

B. **The intermediate layer** consists of the long flexor muscles (fig. 71). **The flexor digitorum longus** (FDL) crosses the flexor hallucis longus (FHL) from below as they emerge from under the sustentaculum tali and receives from it a strong tendinous slip (9). It then divides into four tendons for the four lesser toes. The lumbricals (fig. 72) take origin from two contiguous tendons, except the first lumbrical (4). The obliquity of the pull of these tendons is compensated by a flat muscle which runs along the long axis of the sole of the foot: it arises from the posteromedial and posterolateral tubercles of the calcaneus and is inserted into the lateral border of the tendon for the little toe (fig. 71). By contracting simultaneously, this muscle—the flexor digitorum accessorius (FDA)—decreases the obliquity of these tendons relative to the axis of the foot.

The **flexor hallucis longus** (FHL, figs. 69 and 71) runs in a groove between the two sesamoid bones embedded in the flexor hallucis brevis and gains insertion into the distal phalanx of the big toe.

C. **The superficial layer** consists of a single muscle (fig. 70), which lies in the middle plantar compartment. It is **the flexor digitorum brevis** (FDB), which arises from the posteromedial and posterolateral tubercles of the calcaneus and is inserted by tendon into the middle phalanges of the four lesser toes. It is analogous to the flexor digitorum sublimis of the hand: its tendons are '*perforated*' (fig. 72) and are inserted into the middle phalanx which they flex.

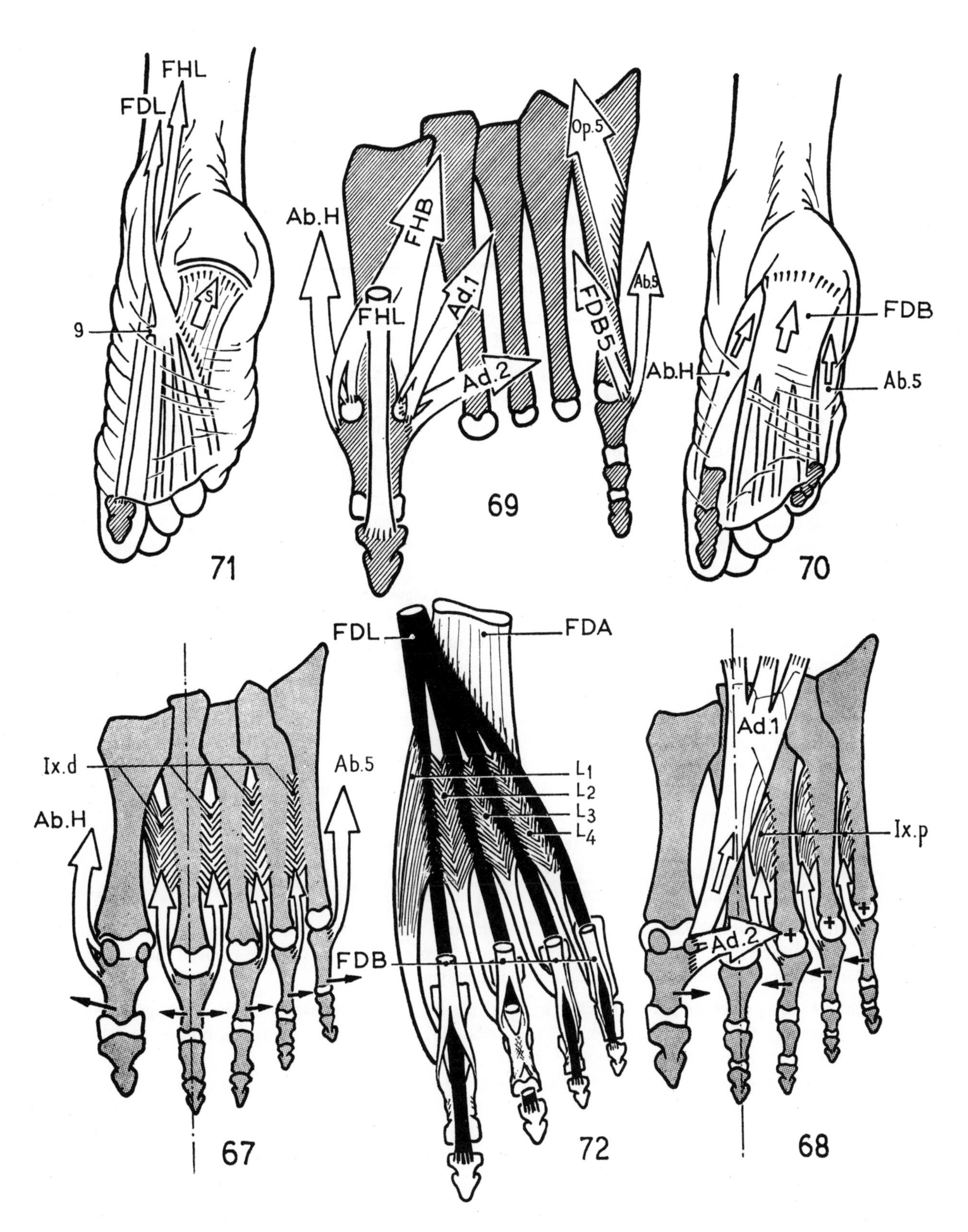
FHL
FDL
9
S
71
Ab.H
FHB
FHL
Ad.1
Ad.2
Op.5
FDB5
Ab.5
69
FDB
Ab.H
Ab.5
70
Ix.d
Ab.5
Ab.H
67
FDL
FDA
L1
L2
L3
L4
FDB
72
Ad.1
Ix.p
Ad.2
68

THE FIBROUS TUNNELS OF THE DORSAL AND PLANTAR ASPECTS OF THE FOOT

The **inferior extensor retinaculum** (fig. 73) keeps the four dorsal tendons of the foot pressed against the tarsal bones at the level of the concavity of the instep: thus it acts as the hinge of a pulley for these tendons whatever the degree of flexion of the ankle. It arises from the superior surface of the calcaneus in front of the sulcus calcanei and soon divides into two diverging bands:

an **inferior band** (a), which blends with the deep fascia of the medial border of the foot;

a **superior band** (b) which is attached to the anterior margin of the tibia near the medial malleolus. This band consists in turn of *distinct lamellae*:

medially, the deep and superficial lamellae embrace the tendon of the tibialis anterior (TA) which is invested in a synovial sheath starting two fingers' breadth proximal to the retinaculum;

laterally these two lamellae form two separate fibrous loops: the medial loop lodges the tendon of the extensor hallucis longus (EHL), invested in a synovial sheath which barely overlaps the retinaculum proximally; the lateral loop contains the tendons of extensor digitorum longus (EDL) and of the peroneus tertius (P.T.), invested in a common synovial sheath which starts slightly above the retinaculum.

All the other tendons pass through tunnels which lie behind the malleoli.

Posterior to the lateral malleolus (fig. 74) the tendons of the peroneus brevis (PB) and of the peroneus longus (PL) run in an osteofibrous tunnel (1) bounded by the fibula and the superior peroneal retinaculum. These tendons are parallel, with the former lying posterior and inferior to the latter. They bend sharply anteriorly below the malleolar tip and are tethered in two osteofibrous tunnels (3 and 4), bounded by the lateral aspect of the calcaneus, the peroneal tubercle (5) and the inferior peroneal retinaculum. At this point their common synovial sheath divides into two separate sheaths. This peroneus brevis (PB) is inserted into the lateral tubercle of the base of the M_V and the base of the M_{IV}. A small segment (7) of this tendon has been resected to display the tendon of the peroneus longus as it changes its direction and enters the groove on the undersurface of the cuboid. The peroneus longus, invested by a new synovial sheath, then runs on the plantar aspect of the foot (fig. 75) through an osteofibrous tunnel constituted superiorly by the tarsal bones and inferiorly by the superficial fibres of the long plantar ligament (fig. 75: deep fibres, 8), which runs from the calcaneus (9) to the cuboid and thence to the bases of all the metatarsals (x), and by fibres of the terminal expansion of the tibialis posterior tendon (TP). The tendon of peroneus longus is inserted mainly into the base of the first metatarsal (11) but also by expansions into the second metatarsal and the medial cuneiform. As it enters the plantar tunnel it is associated almost regularly with a sesamoid bone (12) which allows the tendon to alter its direction more easily.

Therefore the plantar aspect of the foot is carpeted by three sets of fibrous expansions:

the longitudinal fibres of the long plantar ligament;

the fibres of the tendon of the peroneus longus running obliquely anteriorly and medially;

the fibrous expansions of the tibialis posterior tendon which run obliquely anteriorly and laterally to all the tarsal and metatarsal bones, except the last two metatarsals.

Posterior to the medial malleolus (fig. 76) run three tendons which are contained in an osteofibrous tunnel, formed by the tibia and the flexor retinaculum, and are invested in separate synovial sheaths. These tendons are arranged anteroposteriorly and mediolaterally as follows:

The **tibialis posterior** (TP) runs close to the malleolus and bends sharply anteriorly at its tip to gain insertion into the tuberosity of the navicular bone (14) while sending numerous fibrous slips to the plantar aspect of the tarsal and metatarsal bones (10).

The **flexor digitorum longus** (FDL) runs close to the tibialis posterior and along the inner surface of the sustentaculum tali (15; see also fig. 78). It then crosses the deep surface (16) of the flexor hallucis longus.

The **flexor hallucis longus** (FHL) crosses between the posteromedial and posterolateral tubercles (17) of the talus (p. 144) and then below the sustentaculum tali (18; see also fig. 78). Therefore it changes its direction twice.

Two coronal sections (right foot), taken at levels A and B given in figures 74 and 76, illustrate the arrangement of these tendons and of their synovial sheaths in the retromalleolar tunnels: section A (fig. 77) is taken through the malleoli; section B (fig. 78) is more anterior and runs through the sustentaculum tali and the peroneal tubercle.

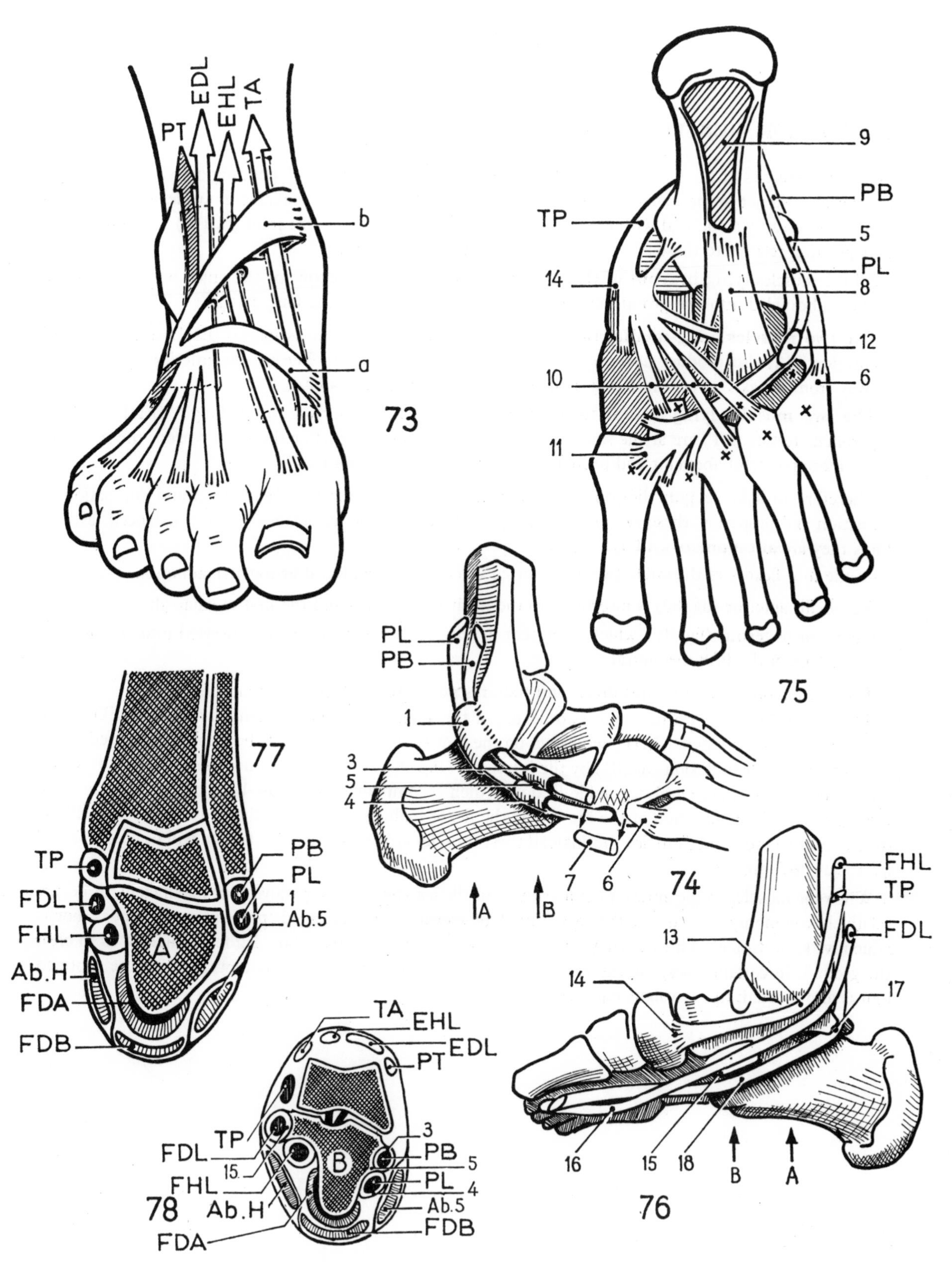
PT
EDL
EHL
TA
b
a
73
9
PB
TP
5
PL
14
8
12
10
6
11
75
PL
PB
1
3
5
4
7
6
74
A
B
77
TP
PB
PL
FDL
1
Ab.5
FHL
A
Ab.H
FDA
FDB
FHL
TP
FDL
13
14
17
16
15
18
B
A
76
TA
EHL
EDL
PT
TP
FDL
15.
3
PB
5
B
FHL
PL
4
78
Ab.H
Ab.5
FDA
FDB

THE FLEXOR MUSCLES OF THE ANKLE

The motor muscles of the foot are either flexors or extensors of the ankle, depending on whether they run anterior or posterior to the transverse axis xx′ of the ankle. Besides, these movements of flexion and extension are necessarily associated with movements around the other axes of the joint (p. 138) depending upon their position relative to those axes (fig. 79, according to Ombredanne).

All the muscles **lying in front of the transverse axis xx′ are flexors** of the ankle but these can be further subdivided into two groups according to their relationship to the long axis zz′:

The two muscles lying medial to this axis, i.e. the extensor hallucis longus (EHL) and the tibialis anterior (TA) also produce *abduction* and *supination* simultaneously. The tibialis anterior, lying farther away from the axis zz′, is more powerful as an adductor and supinator.

The two muscles lying lateral to the axis zz′, i.e. the extensor digitorum longus (EDL) and the peroneus tertius (PT) are at the same time *abductors* and *pronators*. For the same reason, the peroneus is a more powerful abductor and pronator than the extensor digitorum.

Therefore to achieve pure flexion of the ankle, without any associated adduction and supination or abduction and pronation, these two muscle groups must contract simultaneously and in balanced fashion. Thus they are **antagonists and synergists.**

Of the four flexors of the ankle two are inserted directly into the tarsal or metatarsal bones:

the tibialis anterior (fig. 80) is inserted into the medial cuneiform and the first metatarsal;

the peroneus tertius (fig. 81), which is present in only 90 per cent. of cases is inserted into the dorsum of the base of the fifth metatarsal.

Their action on the foot is thus direct and requires no assistance from other muscles.

This is not the case with the other two flexors of the ankle: the extensor digitorum longus (EDL) and the extensor hallucis longus (EHL) which act on the foot via the toes. Thus if the toes are stabilised in the straight position or in flexion (fig. 81) by the interossei (Ix), the extensor digitorum longus flexes the ankle, but, if the interossei are paralysed ankle flexion is then accompanied by a claw-like deformity of the toes. Likewise (fig. 80) stabilisation of the big toe by the flexor hallucis longus and the abductor hallucis allows the extensor hallucis longus to flex the ankle; if these muscles are paralysed ankle flexion will be accompanied by a 'claw toe' (fig. 83).

When the muscles of the anterior compartment of the leg (fig. 82) are paralysed or insufficient, as occurs relatively commonly, the tips of the toes cannot be elevated: this is called 'pes equinus' (equus = horse, which walks on tiptoe). Thus, during walking, the patient must lift the whole leg fairly high so as to clear the ground. In certain cases, the extensor digitorum longus retains some of its power so that the dropped foot is also everted: this is valgus equinus (fig. 85).

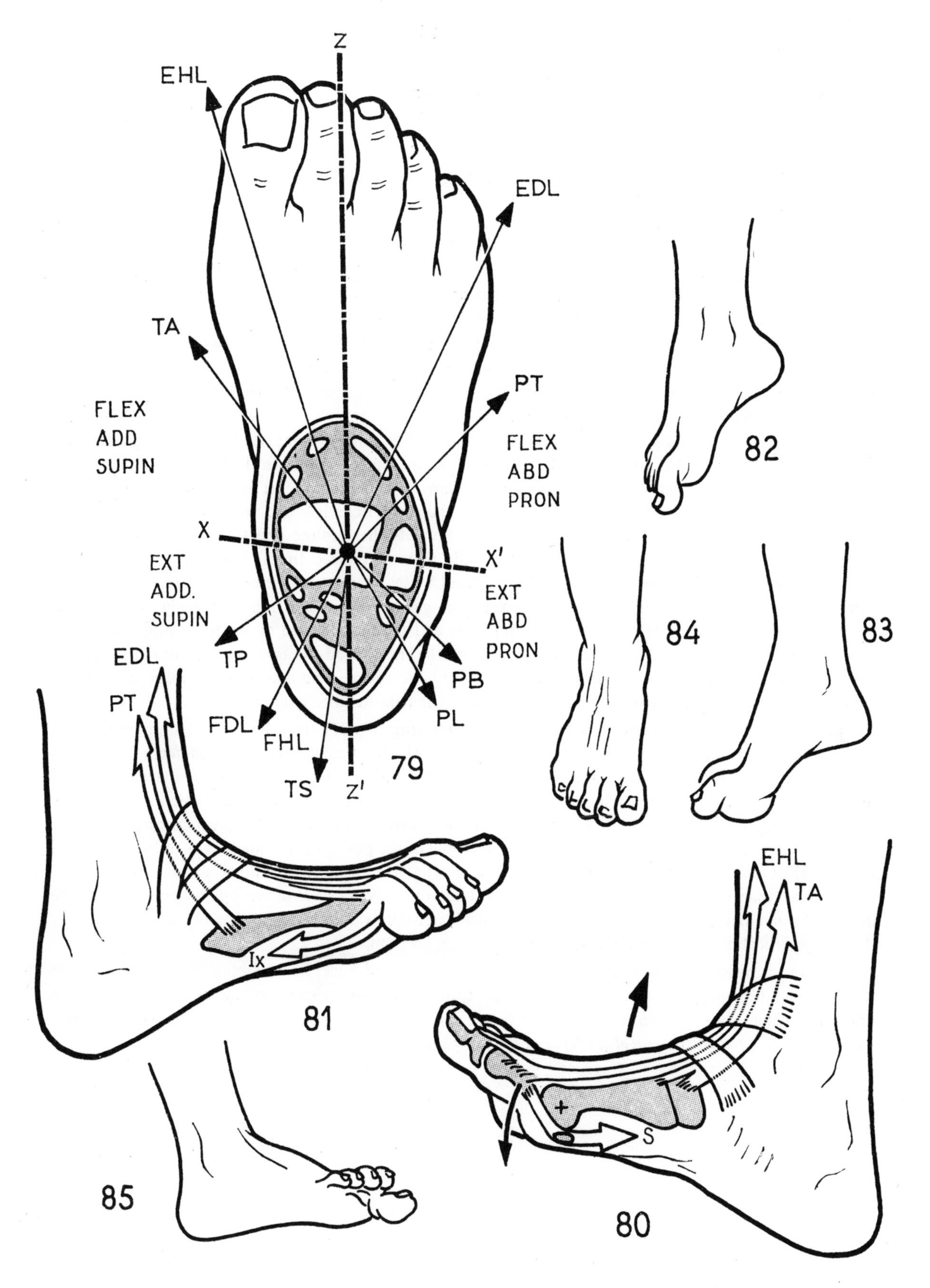

Z
EHL
EDL
TA
PT
FLEX
ADD
SUPIN
FLEX
ABD
PRON
X
X'
EXT
ADD.
SUPIN
EXT
ABD
PRON
TP
PB
FDL
PL
FHL
TS
Z'
79
EDL
PT
Ix
81
82
84
83
EHL
TA
+
S
80
85

THE EXTENSOR MUSCLES OF THE ANKLE: THE TRICEPS SURAE

The extensors of the ankle all lie *behind the axis xx′ of flexion and extension* (fig. 79). Theoretically there are six extensors of the ankle (discounting the plantaris which is negligible). In practice, however, only the triceps surae (gastrocnemius and soleus) are efficient extensors, constituting together one of the most powerful muscles of the body after the gluteus maximus and the quadriceps femoris. On the other hand, its axial position relative to zz′ makes it primarily an extensor. ·

This muscle group consists of **three muscle bellies** (fig. 86) which are inserted by a common tendon—*the Achilles tendon* (1)—into the posterior aspect of the calcaneus (p. 188). Of these three muscle bellies only one is monoarticular, the **soleus** (2), which arises from the tibia, the fibula and a fibrous band stretching between these two bones (3). It is deeply situated—seen here through the gastrocnemius—and surfaces only at the distal end on either side of the Achilles tendon. The other two muscle bellies—the **gastrocnemius**—are biarticular. **The lateral head** (3) arises from an impression above the lateral femoral condyle and from the 'condylar plate' which occasionally contains a sesamoid bone. The **medial head** (5) takes origin from the popliteal surface of the femur above the medial condyle and from the medial 'condylar plate'. These two muscle bellies converge inferiorly towards the midline and form the lower V of the diamond-shaped **popliteal fossa** (10). On either side they are flanked by the hamstring muscles which diverge proximally to form the upper V of the popliteal fossa; laterally by the biceps (6), medially by the sartorius, gracilis and semitendinosus (7). Between the gastrocnemius and the hamstrings *intervene two synovial bursae*: one bursa between the semitendinosus and the medial head of gastrocnemius (8), always present; and the other bursa (9) between the biceps and the lateral head, occasionally present. These bursae can give rise to popliteal cysts. The gastrocnemius and the soleus terminate in a complex aponeurosis (described in anatomical textbooks) which gives rise to the true Achilles tendon.

These three muscles **show unequal degrees of shortening** (fig. 87): the soleus (Cs) 44 mm. the gastrocnemius (Cg) 39 mm. This explains why the efficiency of the biarticular gastrocnemius *depends closely on the degree of flexion of the knee* (fig. 88): as the knee is fully flexed or fully extended the displacement of the origin of the muscle produces a relative lengthening or shortening of the muscle (e), which is equal to or exceeds its length of contraction. Thus when the knee is extended (fig. 89) the gastrocnemius passively stretched works at its best advantage and this allows some of the power of the quadriceps to be transferred to the ankle. On the other hand, when the knee is flexed (fig. 91) the gastrocnemius is maximally slackened (e is greater than Cg) and loses all its efficiency. Thus only *the soleus is active* but its power would be inadequate in walking, riding or jumping unless knee extension was an essential part of the process. Note that the gastrocnemius is not a knee flexor in spite of its position.

Any movement leading to simultaneous extension of ankle and knee, i.e. climbing (fig. 90) or running (figs. 92 and 93) promotes the action of the gastrocnemius. **The triceps surae achieves maximal efficiency** when, starting from the position of flexed ankle-extended knee, it contracts to extend the ankle (fig. 93) and to provide the *propulsive force in the last phase of the step* (Vol. IV).

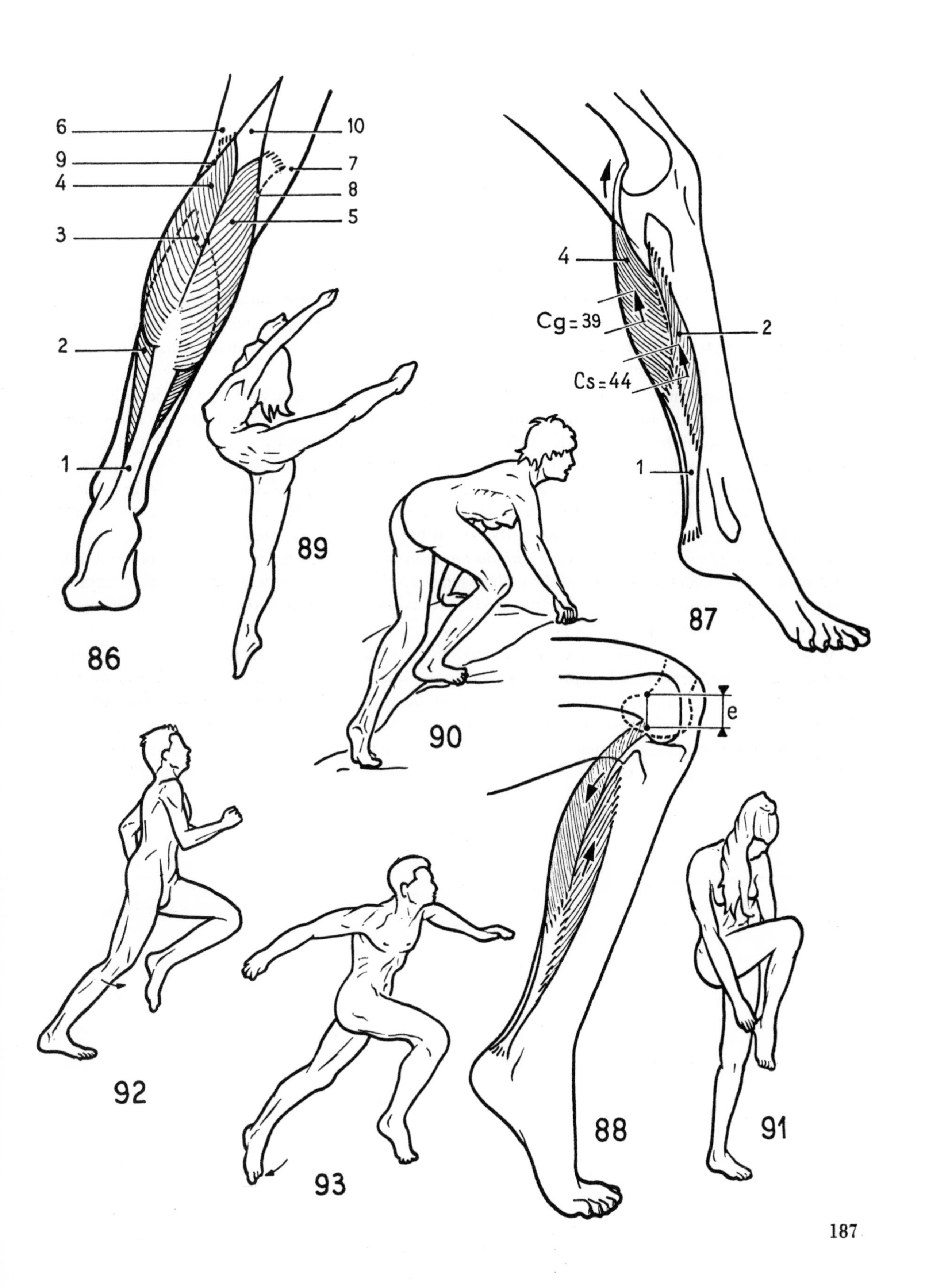
6
9
4
3
2
1
10
7
8
5
86
89
4
Cg=39
Cs=44
2
1
87
90
e
92
93
88
91

The force of the Achilles tendon is applied to the posterior surface of the calcaneus (fig. 94) along a line which forms a *wide angle* with its lever arm AO. When this force is resolved into two vectors, it is found that the effective vector T_1 i.e. perpendicular to the lever arm is much greater than the centripetal vector t_2. Therefore *the muscle works at a high mechanical advantage.*

The effective component t_1 is always greater than t_2 whatever the degree of flexion or extension of the ankle. This is due to the **mode of insertion** of the tendon (fig. 95): it is inserted into the lower part of the posterior surface of the calcaneus (K) while it is separated from the upper part by a bursa. The muscular pull therefore is applied not at the point of insertion (K) but at the point of contact (A) of the tendon with the posterior surface of the bone. With the ankle flexed (a, fig. 95) this point A lies relatively far up on the posterior surface of the calcaneus. With the ankle extended (b, fig. 95) the tendon moves away from the bone and its point of contact A′ now lies farther down but *the direction of the lever arm A′O still remains clearly horizontal, forming a constant angle with the line of the tendon.* This mode of insertion of the Achilles tendon allows the tendon to 'uncoil' on the pulley segment formed by the posterior surface of the calcaneus and this **increases its efficiency during ankle extension**; it resembles the insertion of the triceps brachii into the olecranon process (Vol. I).

When the triceps surae contracts maximally (fig. 96), the movement of extension is associated with a **movement of adduction and supination** so that the sole of the foot faces posteriorly and medially (arrow). This is due to the fact that the **triceps surae acts on the ankle joint through the subtalar joint** (fig. 97). It mobilises these joints in succession (fig. 98): first of all it extends the ankle through 30° around the transverse axis xx′; then acting at the subtalar joint it tilts the calcaneus about the axis of Henke (mn) so that the foot is adducted by 13° and supinated by 12° (Biesalski and Mayer, 1916).

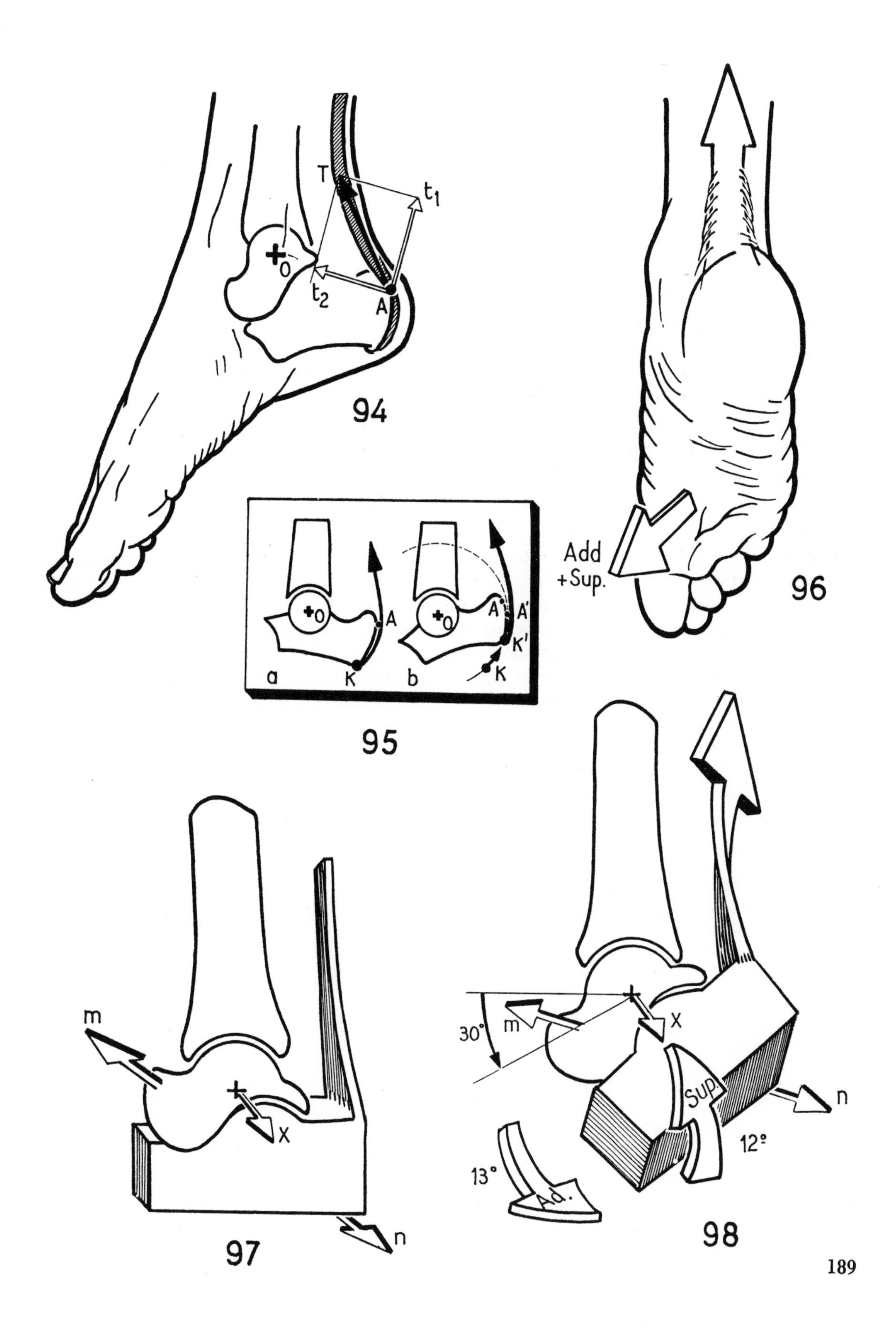
T
t1
+o
t2
A
94
+o
A
a
K
+o
A
A'
K'
b
K
95
Add
+Sup.
96
m
X
97
n
30°
m
X
Sup.
n
12°
13°
Ad.
98

THE OTHER EXTENSOR MUSCLES OF THE ANKLE

All the muscles running posterior to the transverse axis XX′ of flexion and extension (fig. 99) are extensors of the ankle. In addition to the triceps surae (T), **five other muscles** extend the ankle: the plantaris (not described here) is so weak as to be negligible and it is only important in providing a ready tendon for transplantation; unfortunately it is not always present.

Laterally (fig. 100), the peroneus brevis (PB) and the peroneus longus (PL), lying lateral to the long axis zz′ of the joint (fig. 79) simultaneously produce *abduction* and *pronation* (p. 192).

Medially (fig. 101), the tibialis posterior (TP), the flexor digitorum longus (FDL) and the flexor hallucis longus (FHL) lie medial to the axis zz′ (fig. 79) and so simultaneously produce *adduction* and *supination* (p. 194).

Pure extension can result only from balanced action of these lateral and medial muscles i.e. **synergists and antagonists.**

However, the extensor action of these muscles which can be called **accessory extensors,** is relatively slight compared with that of the triceps surae (fig. 102). In effect, the power of the triceps surae is equivalent to 6·5 kg. weight while the total power of these accessory extensors (f) is equivalent to 0·5 kg. weight, i.e. *one fourteenth of the total power available for extension.* The power of a muscle is proportional to its cross-sectional area and its length of contraction (Vol. IV) and so can be represented diagrammatically by a volume whose base is its cross-sectional area and whose height is its length of contraction. The soleus (sol), with cross-sectional area 20 cm^2 and contraction length 44 mm. is less powerful (880) than the gastrocnemius (897) with cross-sectional area 23 cm^2 and contraction length 39 mm. On the other hand, the power of the peroneal muscles (striped cube) represents *one half* of the total power of the accessory extensors. The peroneus longus is itself in turn twice as powerful as the peroneus brevis.

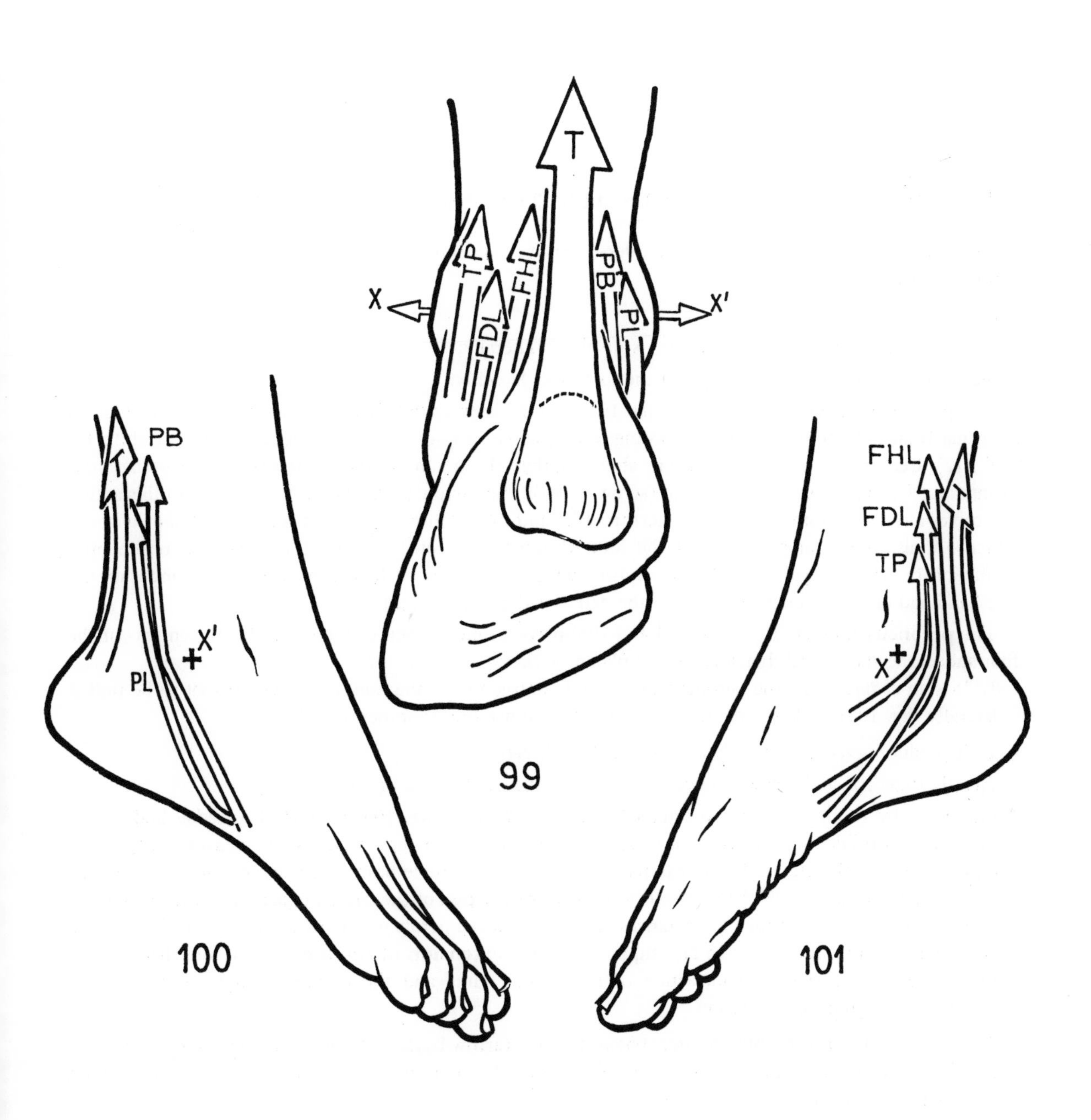
T
TP
FHL
PB
X
X'
FDL
PL
PB
T
X'
PL
FHL
FDL
TP
T
X
99
100
101

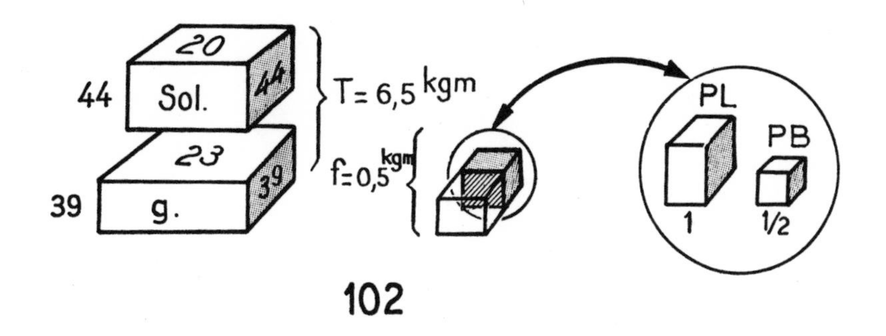
20
44
Sol.
44
T= 6,5 kgm
23
39
g.
39
f=0,5 kgm
PL
PB
1
1/2
102

THE ABDUCTOR-PRONATOR MUSCLES: THE PERONEI

These muscles run posterior to the transverse axis xx′ and outside the long axis zz′ and so produce simultaneously (fig. 103):

extension (arrow 1);

abduction (arrow 2) so that the axis zz′ is displaced laterally;

pronation (arrow 3) so that the sole of the foot faces laterally.

The **peroneus brevis** (PB), inserted (fig. 104) into the lateral tubercle of the base of the fifth metatarsal is primarily an abductor of the foot: according to Duchenne de Boulogne it is in fact the only pure abductor (fig. 112). Certainly, it is a more efficient abductor than the peroneus longus. It also produces (fig. 105) pronation of the anterior half of the foot (arrow 3) by lifting (arrow a) the lateral metatarsals: in this action it receives assistance from the peroneus tertius (PT) and the extensor digitorum longus (not shown here), which are also abductor-pronators of the foot while simultaneously flexing the ankle. Therefore pure abduction-pronation, results from the *synergistic-antagonistic action* of the peronei brevis and longus on the one hand and of the peroneus brevis and the extensor digitorum longus on the other.

The **peroneus longus** (PL) (figs. 104 and 106) plays a fundamental part both in the movements of the foot and in the statics and dynamics of the plantar arches:

1. It is an *abductor* like the peroneus brevis and contracture of the muscle causes the foot to be pulled laterally (fig. 108) while the medial malleolus sticks out more prominently.

2. It produces *extension* directly and especially indirectly:

   directly (figs. 105 and 106), by lowering (arrow 6) the head of the first metatarsal;

   indirectly, by pulling the first metatarsal laterally (arrow 5, fig. 106) so that the medial and lateral metatarsals form one solid piece, as it were. Now (fig. 107) the triceps surae, as an extensor, acts directly on only the lateral metatarsals (shown diagrammatically as a single beam): thus by ‘coupling’ the medial with the lateral metatarsals the peroneus longus allows the pull of the triceps to act on all the metatarsals at one time. This is confirmed by instances of paralysis of the peroneus longus where only the lateral arch is extended so that the foot is in fact supinated. **Pure extension of the foot** is therefore the result of the *synergistic-anatagonistic* contraction of the triceps surae and the peroneus longus: synergistic in extension, antagonistic in rotation.

3. It is in effect a *pronator* muscle (fig. 105) as it lowers (arrow b) the head of the first metatarsal when the foot is clear of the ground. Pronation (arrow 3) results from elevation of the lateral arch (a) along with lowering of the medial arch (b).

It will become apparent later (p. 204) how the peroneus longus accentuates the curvature of the three arches of the foot and constitutes their main muscular support.

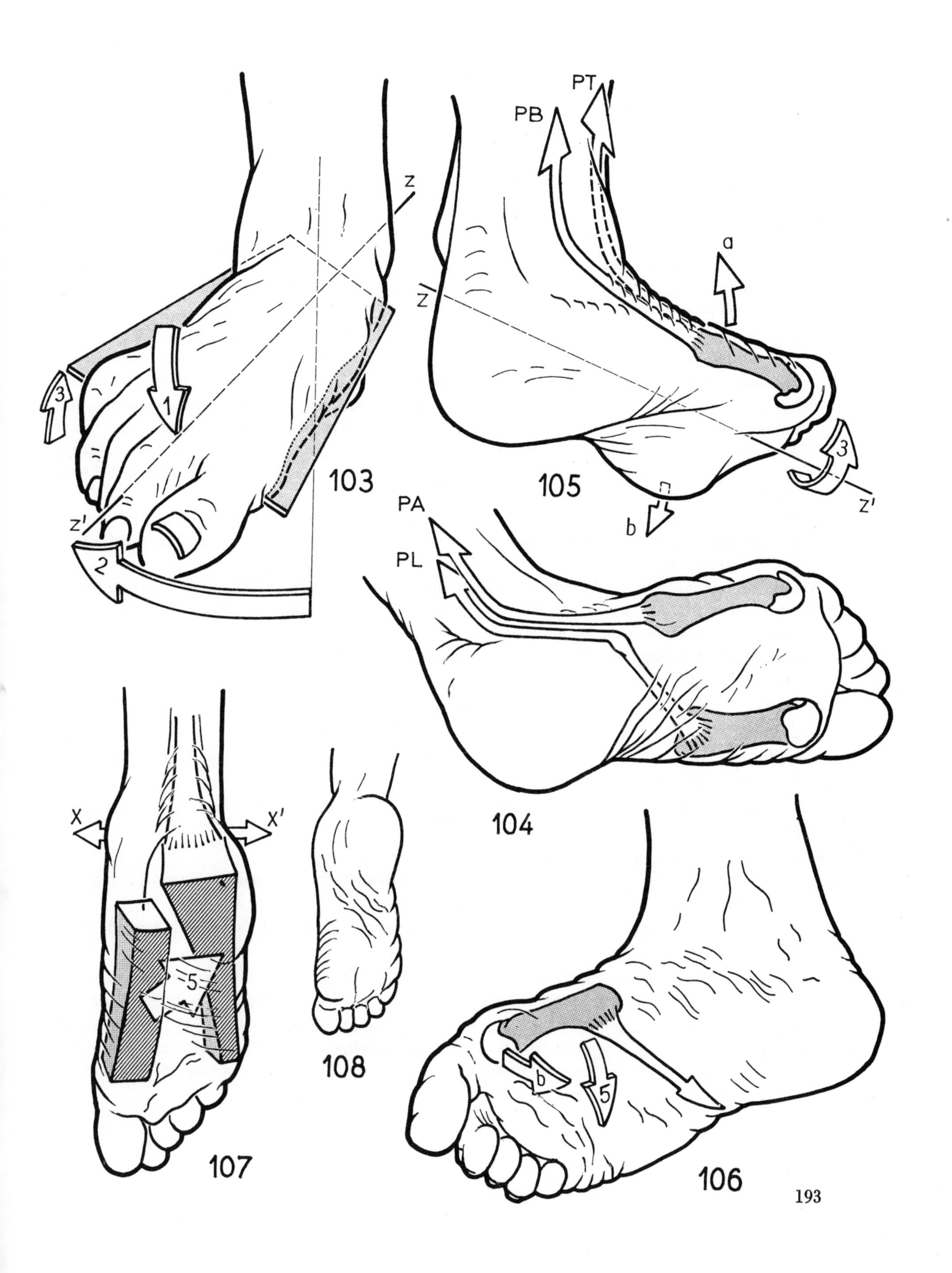
Z
Z'
1
2
3
103
PT
PB
a
Z
3
Z'
b
105
PA
PL
104
X
X'
5
107
108
b
5
106

THE ADDUCTOR-SUPINATOR MUSCLES: THE TIBIALIS MUSCLES

The three muscles, lying posterior to the medial malleolus, run posterior to the axis xx′ and inside the axis xx′ (fig. 79) and so produce simultaneously (fig. 109):

extension (arrow 1);

adduction (arrow 2), so that the axis zz′ is displaced medially;

supination (arrow 3) so that the sole of the foot faces medially.

The **tibialis posterior** (TP), the most important of these three muscles, is inserted (fig. 110) into the tubercle of the navicular bone (shaded). As it crosses the ankle and the subtalar and transverse tarsal joints it acts simultaneously on all three:

By pulling the navicular medially (fig. 111) it is a **very powerful adductor** (according to Duchenne de Boulogne it is more of an adductor than supinator). Thus it is a *direct antagonist of the peroneus brevis* (PB), which draws the anterior part of the foot laterally (fig. 112) by acting on the fifth metatarsal.

As a result of its plantar attachments to the tarsal and metatarsal bones (fig. 75) it produces **supination** and plays a vital role in supporting and orientating the plantar arches (p. 204). Congenital absence of these plantar attachments has been blamed as one of the causes of the pes planus valgus. The total range of supination is 52° with 34° occurring at the talocalcanean joint and 18° at the transverse tarsal joint (Biesalski and Mayer).

It is an extensor (fig. 113) not only of the ankle (arrow a) but also of the transverse tarsal joint by lowering the navicular (arrow b): the ankle movement is continued by the movement of the anterior half of the foot (p. 141, fig. 5).

The tibialis anterior (TA) and the extensor hallucis longus (EHL) pass (fig. 114) *anterior to* the transverse axis xx′ and medial to the long axis zz′ (fig. 79): they therefore produce simultaneously *flexion* of the ankle and *adduction* and *supination* of the foot.

The **tibialis anterior** (fig. 109) is more efficient as a supinator than an adductor. It acts by elevating all the structures of the medial arch (fig. 113).

It lifts the base of the first metatarsal on the medial cuneiform (arrow c) so that the metatarsal head is also elevated.

It lifts the medial cuneiform over the navicular (arrow d) and the navicular over the talus (arrow e) before flexing the ankle (arrow f).

It flattens the medial arch by producing supination of the foot and so is *the direct antagonist of the peroneus longus*.

It has a less powerful action as an adductor than the tibialis posterior.

It flexes the ankle and, in conjunction with its *synergist-antagonist* i.e. the tibialis posterior, it produces pure adduction and supination without flexion or extension.

Contracture of the tibialis anterior causes a pes talovarus with flexion deformity of the toes (fig. 115), especially of the big toe.

The **extensor hallucis longus** (fig. 114) is less powerful than the tibialis anterior in producing adduction and supination. It can replace the latter as a flexor of the ankle but it often produces 'clawing' of the big toe.

The power of the supinators (2·82 kg. weight) exceeds that of the pronators (1·16 kg. weight): in the absence of any support the foot spontaneously assumes a position in supination. This imbalance compensates beforehand for the natural tendency of the foot to be pronated (p. 206) when it supports the weight of the body on the ground.

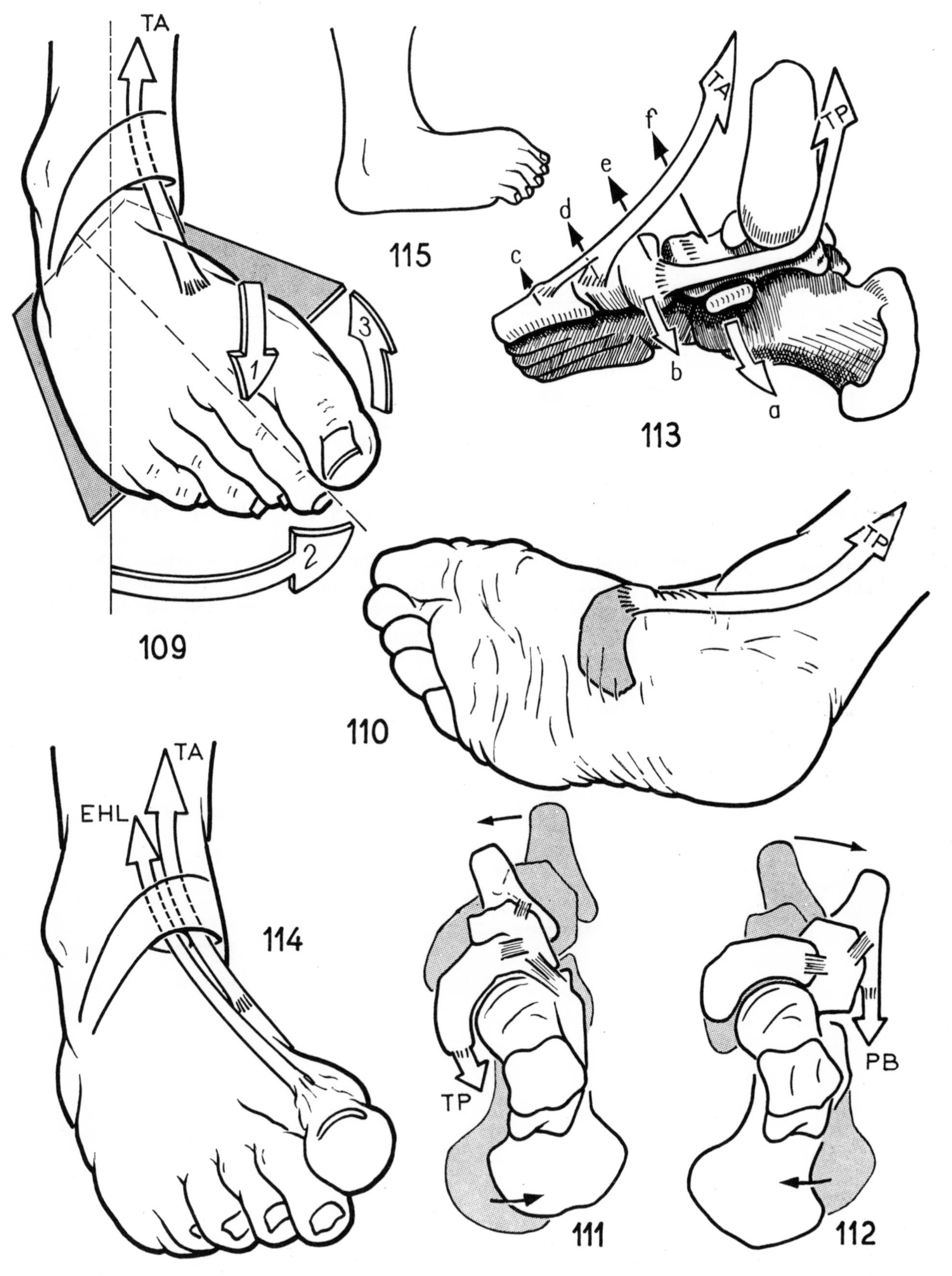

TA
1
2
3
109
115
TA
TP
c
d
e
f
b
a
113
TP
110
TA
EHL
114
TP
111
PB
112

THE PLANTAR VAULT OR THE ARCHES OF THE FOOT

The plantar vault is an **architectural structure** which blends all the elements of the foot—joints, ligaments and muscles—into a unified system. Thanks to its changes of curvature and its elasticity, the vault can adapt itself to unevenness of the ground and can transmit to the ground the forces exerted by the weight of the body and its movements. This it achieves with the best mechanical advantage under the most varied conditions. The plantar vault acts as a **shock-absorber** essential for the flexibility of the gait. Any pathological conditions, which exaggerate or flatten its curvatures, interfere seriously with the support of the body on the ground and necessarily with running, walking and the maintenance of the erect posture.

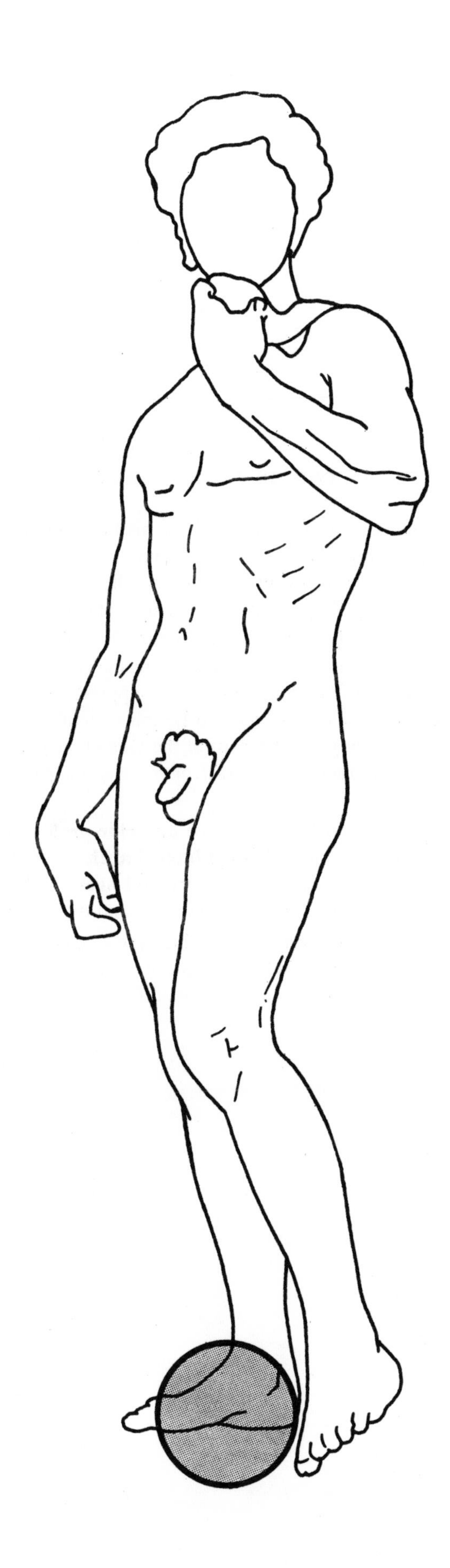

GENERAL ARCHITECTURE OF THE PLANTAR VAULT

Viewed as a whole, the plantar vault can be compared with an **architectural vault supported by three arches** (fig. 1). It rests on the ground at three points A, B and C (fig. 2) which lie at the corners of an equilateral triangle (fig. 2). Between two consecutive supports AB, BC or CA stretches an arch which constitutes one of the sides of the vault. The weight of the vault is applied (fig. 3) at the *keystone* (arrow) and is distributed to the supports A and B (known as the *abutment piers* of the arch) by the two buttresses.

The **plantar vault** (fig. 4: seen from inside; the structures are shown as transparent) does not form an equilateral triangle but, as it contains three arches and three supports, its structure is comparable. Its supports (fig. 5: seen from above; the foot is assumed to be transparent) lie within the zone of contact with the ground or the **footprint** (striped). They consist of the *head of the first metatarsal* (A), *the head of the fifth metatarsal* (B) and the *posteromedial and lateral tubercles of the calcaneus* (C). Each support is shared by two adjacent arches.

Between the two *anterior* supports A and B stretches the **anterior arch** which is the shortest and the lowest. Between the two *lateral* supports B and C lies the **lateral arch** of intermediate length and height. Finally between the two *medial* supports C and A lies the **medial arch** the longest and highest and also the most important of the three during static support of the body and during movements.

The shape of the plantar vault (the lower part of fig. 4) therefore resembles that of a jib swollen by the wind. Its top is distinctly displaced posteriorly and the weight of the body is applied on its posterior slope (arrow) at a point (black cross in fig. 5) located at the centre of the instep.

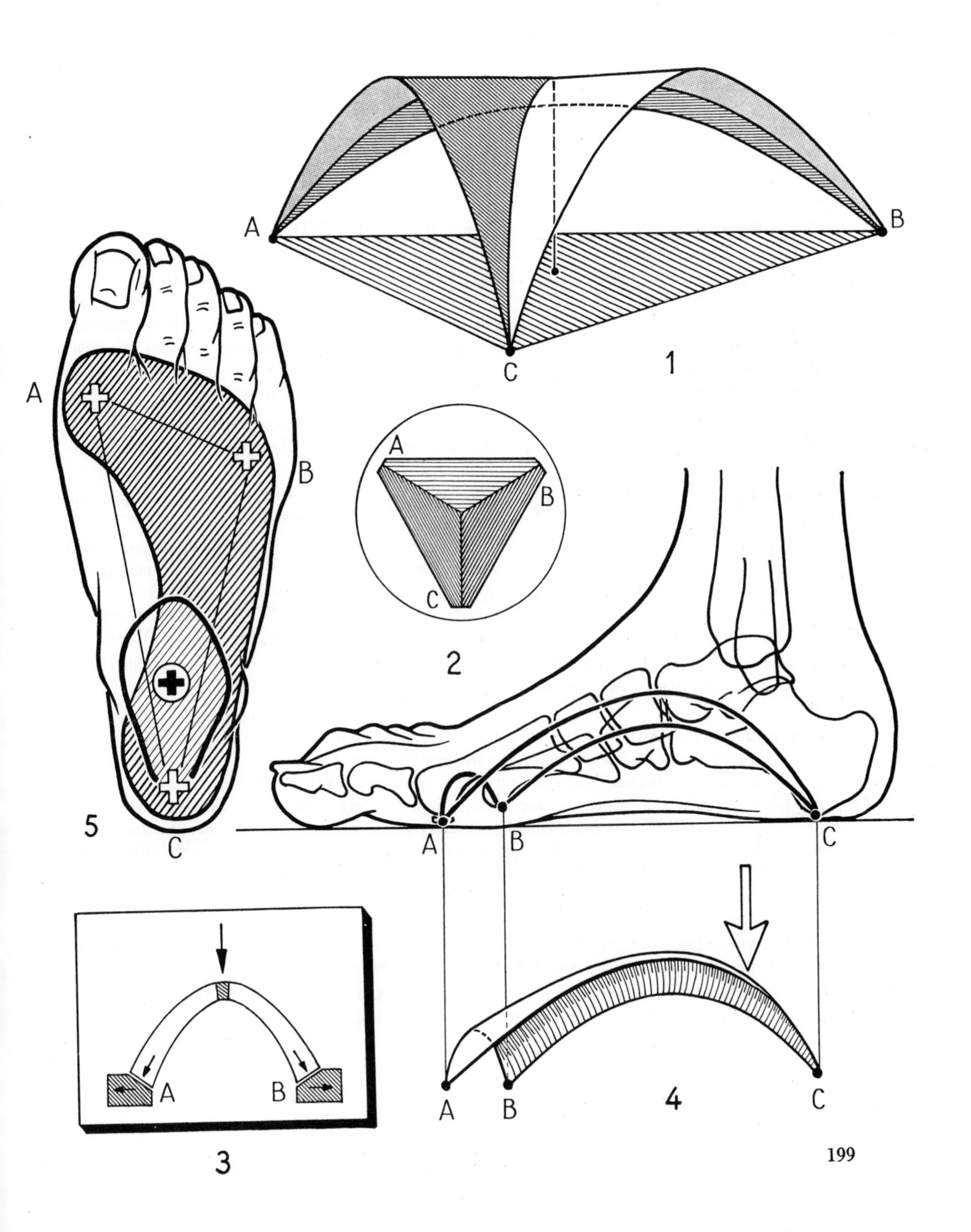
A
B
C
1
A
B
C
2
A
B
C
5
A
B
C
A
B
C
4
A
B
3

THE THREE ARCHES OF THE PLANTAR VAULT: THE MEDIAL ARCH

Between its anterior (A) and posterior (C) supports the medial arch (fig. 6) comprises **five bones** which are as follows anteroposteriorly:

the *first metatarsal* (M_1) touches the ground only by its head (A);
the *medial cuneiform* (C_m), completely clear of the ground;
the *navicular* (nav.), which is the keystone of the arch (striped) and lies 15 to 18 cm. above the ground;
the *talus* (Tal), which receives all the forces transmitted by the leg and transmits them to the vault (fig. 33);
the *calcaneus* (Calc.) which is in contact with the ground only at its posterior extremity.

The transmission of the mechanical forces is reflected (fig. 7) in **the direction of the bony trabeculae.**

The trabeculae arising from the cortex of the anterior surface of the tibia run obliquely inferiorly and posteriorly through the posterior buttress of the arch. They traverse the body of the talus and fan out in the calcaneus to reach the posterior support of the arch.

The trabeculae arising from the cortex of the posterior surface of the tibia run obliquely inferiorly and anteriorly and traverse the neck and head of the talus, the navicular and the anterior buttress, i.e. the medial cuneiform and the metatarsal.

The medial arch maintains its concavity only *with the help of ligaments and muscles* (fig. 6).

Many **plantar ligaments** unite these five bones: cuneometatarsal, cuneonavicular, but especially the *plantar calcaneonavicular ligament* (1) and the *talocalcanean ligament* (2). These resist violent but short-lasting stresses whereas the muscles cope with sustained strains.

The **muscles** join two points, which lie at varying distances along the arch, and span either the whole of the arch or part of it. They therefore act as **tighteners** of the various arches.

The **tibialis posterior** (TP) spans part of the medial arch (fig. 9) near its dome but it plays a vital part. In fact (fig. 8) it pulls back the navicular inferiorly and posteriorly under the head of the talus (circle with circumference in broken line). This relatively trivial shortening of the muscle (e) is associated with a change in direction of the navicular so that the anterior buttress of the arch is lowered. Moreover, its plantar attachments (3, fig. 6) blend with the plantar ligaments and act on the three middle metatarsals.

The **peroneus longus** (PL) also acts on the medial arch and accentuates its curvature (fig. 10) by flexing M_1 on the medial cuneiform and the latter on the navicular (fig. 8); (see also p. 204 for its action on the transverse arch).

The **flexor hallucis longus** (FHL) spans most of the medial arch (fig. 11) and so has a powerful influence on its curvature; in this action it is assisted by the **flexor digitorum longus** (FDL) which crosses it from below (fig. 12). The flexor hallucis longus also acts to stabilise the talus and the calcaneus: as it runs between the two tubercles of the talus it prevents the talus (r) from receding when pushed back by the navicular (white arrow): the talocalcanean interosseous ligament (2) is first stretched and *the talus is restored to its original position anteriorly* by the tendon which propels it forward just as a bowstring propels an arrow. As it runs below the sustentaculum tali (fig. 14) the flexor hallucis longus (by a similar mechanism) re-elevates the anterior half of the calcaneus which receives the vertical force transmitted by the head of the talus (white arrow).

The **abductor hallucis longus** (Ab. HL.) spans the whole medial arch (fig. 15). It is therefore a particularly efficient tightener: it accentuates the curvature of the arch by approximating its two ends.

On the other hand (fig. 16) the two muscles inserted into the convexity of the arch, i.e. the extensor hallucis longus (EHL)—under certain conditions—and the tibialis anterior (TA) reduce its curvature and flatten the arch.

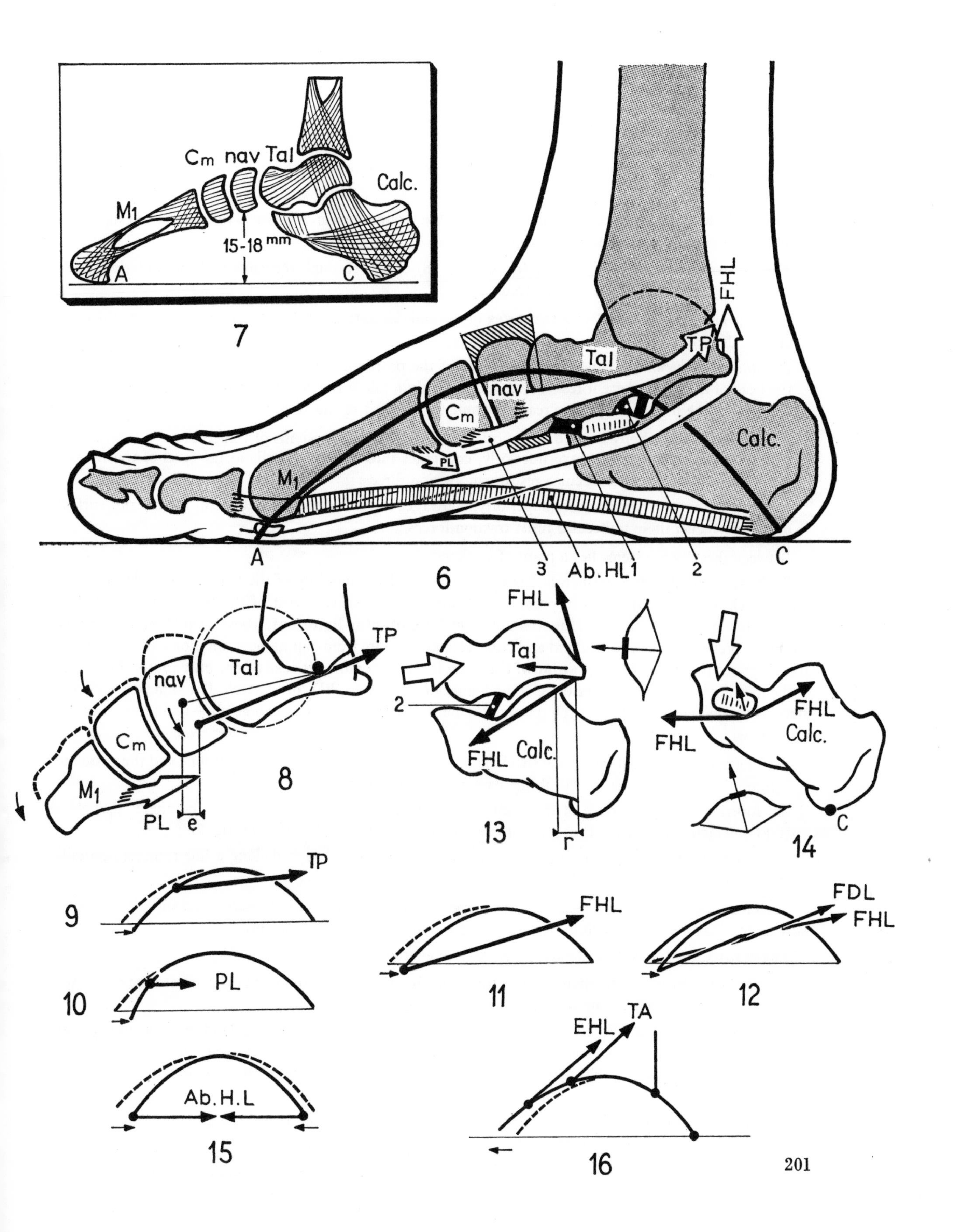

Cm
nav
Tal
Calc.
M1
15-18 mm
A
C
7
FHL
TP
Tal
nav
Cm
PL
M1
Calc.
A
3
Ab. HL1
2
C
6
TP
Tal
nav
Cm
M1
PL
e
8
FHL
Tal
2
FHL
Calc.
r
13
FHL
Calc.
FHL
C
14
TP
9
PL
10
Ab.H.L
15
FHL
11
FDL
FHL
12
EHL
TA
16

THE LATERAL ARCH

This comprises only **three bones** (fig. 17):

the *fifth metatarsal* (M_5) with its head constituting the anterior support (B) of the anterior arch;

the *cuboid* (Cub), completely clear of the ground;

the *calcaneus* (Calc), with its posteromedial and posterolateral tubercles forming the posterior support (C) of the arch.

In contrast with the medial arch, which overhangs the ground, this arch is low (3 to 5 mm.) and establishes contact with the ground by means of the soft tissues.

The transmission of mechanical forces (fig. 18) occurs through the talus and the underlying calcaneus. Two sets of trabeculae are involved (fig. 18):

the *posterior trabeculae* arise from the cortex of the anterior surface of the tibia, and fan out into the body of the calcaneum;

the *anterior trabeculae*, arising from the cortex of the posterior surface of the tibia, first of all traverse the head of the talus where it rests on the sustentaculum tali of the calcaneum and then run through the cuboid and the fifth metatarsal to reach the anterior support of the arch.

In addition to these bony trabeculae the calcaneus has two main systems of trabeculae:
a *superior arcuate system*, concave inferiorly, converges into a dense lamella at the level of the floor of the sinus calcanei; these trabeculae resist compression forces;

an inferior arcuate system, concave superiorly, converges towards the cortical bone of the inferior surface of the calcaneus; these trabeculae resist traction forces.

Between these two systems lies a point of weakness (+).

While the medial arch is eminently flexible as a result of the mobility of the talus on the calcaneus, the lateral arch is much more rigid in order to transmit the propulsive thrust of the triceps surae (fig. 107, p. 107). Its rigidity depends upon the strength of the **long plantar ligament**, whose deep (4) and superficial (5) fibres prevent the calcaneocuboid and cubometatarsal joints from gaping open inferiorly (fig. 19) under the weight of the body. **The keystone of the arch is the anterior process of the calcaneus** (D), which is the meeting point of the posterior (CD) and anterior buttresses (BD) of the arch. If a sufficiently violent stress is applied vertically to the arch through the talus—fall on the foot from a height—two types of injury can follow:

the long plantar ligament resists the shock but the arch snaps at the level of its keystone and the sustentaculum is broken off along a vertical line passing through the point of weakness;

the superior articular surface of the calcaneus is impacted into the bone so that the angle of Boehler (PTD), which is normally obtuse, is flattened out or even reversed to PT′ D (fig. 20);

on the medial side, the medial tubercle of the calcaneus is often detached along a line running sagittally (not shown here).

Such fractures of the calcaneus are not easily reduced as one must not only restore the superior articular surface but also the sustentaculum tali, otherwise the medial arch stays collapsed.

Three muscles act as *active tighteners* of the lateral arch:

the **peroneus brevis** (PB) spans part of the arch (fig. 21) but like the plantar ligament it prevents the joints of the foot from gaping open inferiorly (fig. 22);

the peroneus longus (PL), which runs parallel to the former as far as the cuboid, plays a similar role; but in addition (fig. 23), as it is attached to the peroneal tubercle of the calcaneus (6), it props up the anterior extremity of the calcaneus by its own elasticity just as the flexor hallucis longus does on the medial side;

the **abductor digiti minimi** (Ab. 5) spans the whole length of the lateral arch (fig. 24) and it has a similar action to that of its counterpart, the adductor hallucis longus.

Acting on the convexity of the lateral arch (fig. 25), the peroneus tertius (PT) and the extensor digitorum longus (EDL)—under certain conditions—decrease its curvature. The triceps surae (TS) has a similar effect.

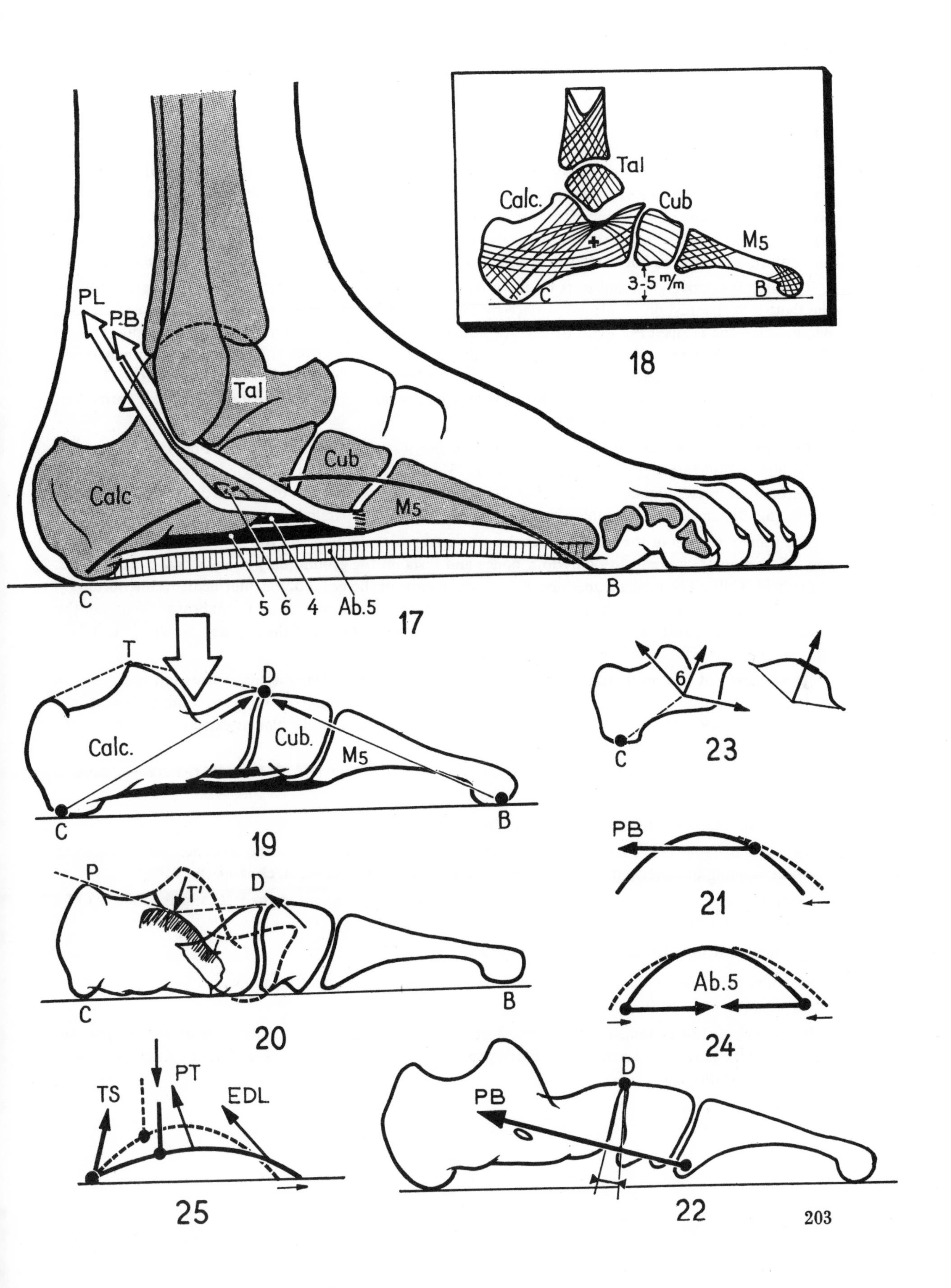
PL
P.B.
Tal
Cub
Calc
M5
C
5
6
4
Ab.5
B
17
Tal
Calc.
Cub
M5
3-5 m/m
C
B
18
T
D
Calc.
Cub.
M5
C
B
19
P
T'
D
C
B
20
6
C
23
PB
21
Ab.5
24
TS
PT
EDL
25
D
PB
22

THE ANTERIOR ARCH AND THE TRANSVERSE CURVATURE OF THE FOOT

The anterior arch (fig. 26, section I) runs from **the head of the first metatarsal** (A), **which rests on two sesamoid bones** and is 6 mm. above the ground, to **the head of the fifth metatarsal** (**B**), which also lies 6 mm. above the ground. It traverses the heads of the intervening metatarsals. The **head of the second metatarsal**, which is the highest above the ground (9 mm.), is the *keystone* of the arch. The head of the third (8·5 mm.) and of the fourth (7 mm.) metatarsals occupy intermediate positions (see also fig. 37).

The arch is relatively flat and rests on the ground via the soft tissues, often called the 'anterior heel' of the foot. It is spanned on its plantar surface by the relatively weak intermetatarsal ligaments and by only one muscle—the **transverse head of the adductor hallucis** (Ad. H). Some fibres of the adductor span the whole length of the arch while others span only part of it running from the head of the first metatarsal, to each of the other metatarsal heads. This muscle is relatively weak and gives way very easily. The arch is often flattened—flat forefoot—or even reversed—the convex forefoot—so that callosities are formed on the lowered metatarsal heads (p. 218).

The *anterior arch is the point of culmination of the five metatarsal rays of the foot.* The first ray (fig. 28) is the highest and forms (Fick) an angle of 18° to 25° with the ground. This angle between the metatarsal and ground decreases regularly, being 15° for the second ray (fig. 29), 10° of the third (fig. 30), 8° for the fourth (fig. 31) and only 5° for the fifth (fig. 32), which is nearly parallel to the ground.

The transverse arch of the foot involves the whole length of the foot. **At the level of the cuneiforms** (fig. 26, section II) it comprises only four bones and rests on the ground only at its lateral extremity, i.e. the cuboid (Cub). The medial cuneiform (C_m) is quite clear of the ground; the intermediate cuneiform (C_i) is the keystone of the arch (striped) and constitutes with the second metatarsal the axis of the foot, i.e. **the crest-tile of the vault**. This arc of a circle is subtended by the tendon of the peroneus longus (PL) which thus exerts a powerful action on the transverse curvature of the foot.

At the level of the navicular and the cuboid (fig. 26, section III) the transverse arch rests only on its lateral extremity, i.e. the cuboid (Cub). The navicular (nav) is slung above the ground and overhangs the medial surface of the cuboid. The curvature of this arch depends on the plantar expansions of the tibialis posterior (TP).

A plantar view of the left foot (assumed to be transparent) shows (fig. 27) how the **transverse arch** is maintained by three muscles successively anteroposteriorly:

the adductor hallucis (Ad. H) which runs transversely;

the peroneus longus (PL), the most important muscle in the dynamics of the foot, acts as a tightening device running obliquely anteriorly and medially: it acts **on the three arches of the foot**;

the plantar expansions of the tibialis posterior (TP), especially important in the statics of the foot, act as a tightener running obliquely anteriorly and laterally.

The **longitudinal curvature** of the foot as a whole depends on:

medially, the *abductor hallucis* (Ab. H) and the flexor hallucis longus (not shown);

laterally, the *abductor digiti minimi* (Ab. 5).

Between these two extreme tighteners the flexor digitorum longus, the flexor digitorum accessorius (not shown) and the flexor digitorum brevis (FDB) maintain the curvature of the three intermediate rays as well as that of the fifth ray.

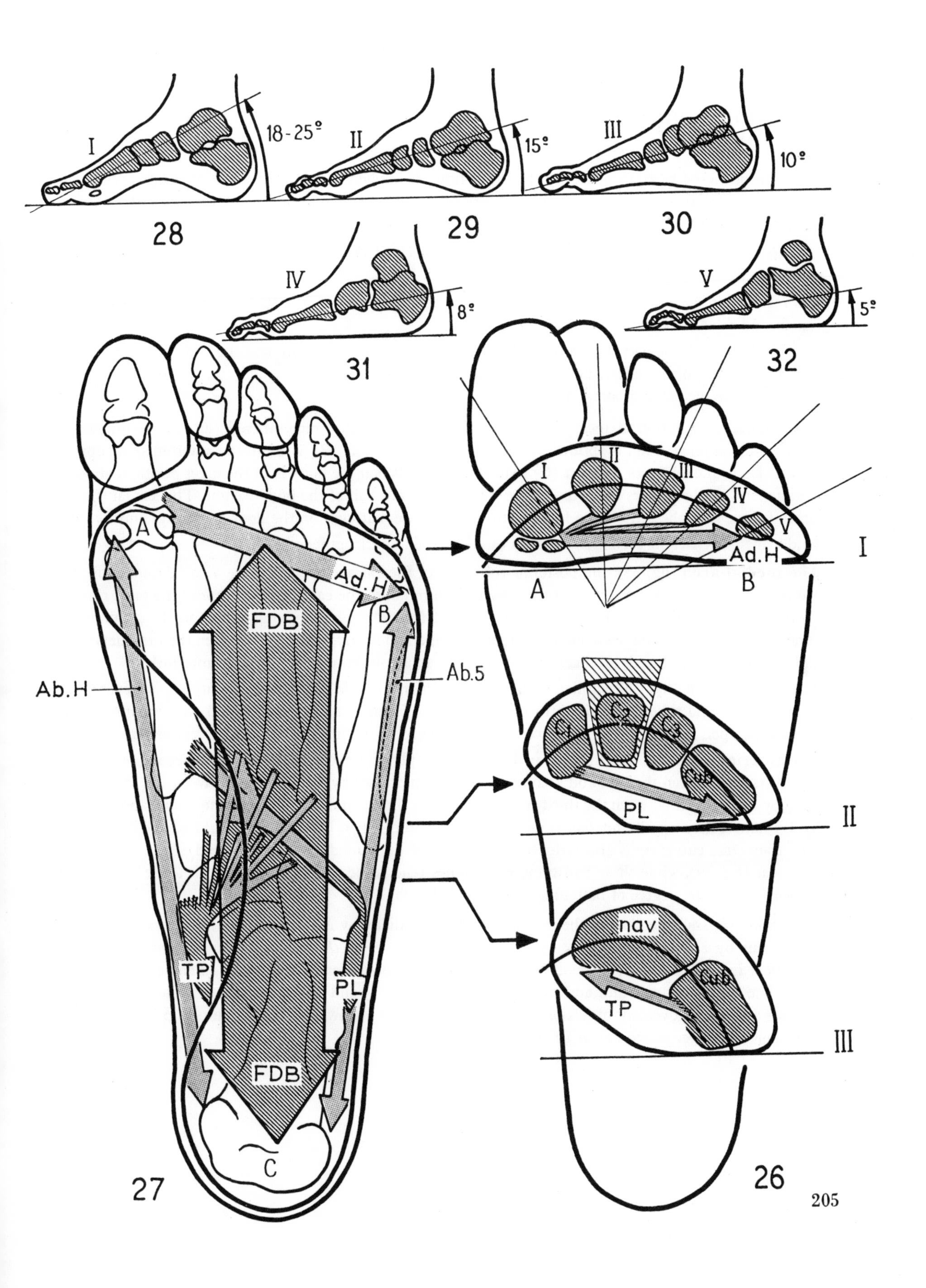

I
18-25°
II
15°
III
10°
28
29
30
IV
8°
V
5°
31
32
A
Ad.H
FDB
B
Ab.H
Ab.5
TP
PL
FDB
C
27
I
II
III
IV
V
Ad.H
I
A
B
C1
C2
C3
Cub
PL
II
nav
Cub
TP
III
26

THE DISTRIBUTION OF STRESSES AND THE STATIC DISTORTIONS OF THE PLANTAR VAULT

The weight of the body, transmitted by the lower limb, is applied through the ankle to the posterior part of the foot (fig. 33) at the level of the trochlear surface of the talus. From there the forces are distributed in three directions towards the supports of the vault (Seitz, 1901):

towards the anterior and medial support (A), via the neck of the talus and the anterior buttress of the medial arch;

towards the anterior and lateral support (B), via the head of the talus, the sustentaculum tali of the calcaneus and the anterior buttress of the lateral arch;

towards the posterior support (C) via the body of the talus, the subtalar joint, and the body of the calcaneus (the bony trabeculae underlying the superior articular surface) i.e. through the common posterior buttress of the medial and lateral arches.

The relative distribution of these forces to each support is easily remembered (fig. 34) as follows: if 6 kg. weight is applied, 1 kg. is applied to the anterolateral support (B), 2 kg. to the anteromedial support (A) and 3 kg. to the posterior support (C) (Morton, 1935). When the body is in the erect position, straight and stationary, the heel bears the brunt of the stress, i.e. about half of the body weight. This explains why, when this force is applied through a fine stiletto heel, plastic materials on the floor are easily 'punched in'.

Under the body weight each arch of the foot is flattened and lengthened.

The medial arch (fig. 35): the posterior tubercles of the calcaneum which are 7 to 10 mm. above the ground, are lowered by 1·5 mm. and the sustentaculum tali of the calcaneus by 4 mm.; the talus recedes on the calcaneus; the navicular rises on the head of the talus while moving nearer to the ground; the cuneonavicular and the cuneometatarsal joints gape open inferiorly; the angle between the first metatarsal and the ground is reduced; the heel recedes and the sesamoid bones move a little anteriorly.

The lateral arch (fig. 36): similar vertical displacements of the calcaneus; the cuboid is lowered by 4 mm., the lateral tubercle of the fifth metatarsal by 3·5 mm.; the calcaneocuboid and cubometatarsal joints gape open inferiorly; the heel recedes and the head of the fifth metatarsal moves forward slightly.

The anterior arch (fig. 37): the arch is flattened and splayed out on either side of the second metatarsal. The distance between the first and second metatarsals increases by 5 mm., that between the second and third by 2 mm., that between the third and the fourth by 4 mm. and that between the fourth and the fifth by 1·5 mm. Therefore the *forefoot is widened* by 12·5 mm. when bearing the weight of the body.

The transverse curvature is also reduced at the level of the cuneiforms (fig. 38) and at the level of the navicular (fig. 39), while these two transverse arches tend to be tilted laterally by an angle x, which is proportional to the degree of flattening of the medial arch.

In addition (fig. 40) the head of the talus is displaced medially by 2 to 6 mm. and the lateral tubercle of the calcaneus by 2 to 4 mm. This leads to *twisting of the foot at the transverse tarsal joint*: the axis of the hindfoot is displaced medially while that of the forefoot moves laterally forming an angle of y with the former. The hindfoot turns into adduction-pronation (arrow 1) and slight extension while the forefoot undergoes a *relative* movement of flexion-abduction-supination (arrow 2). This phenomenon is particularly conspicuous in the *pes planus valgus* (p. 216).

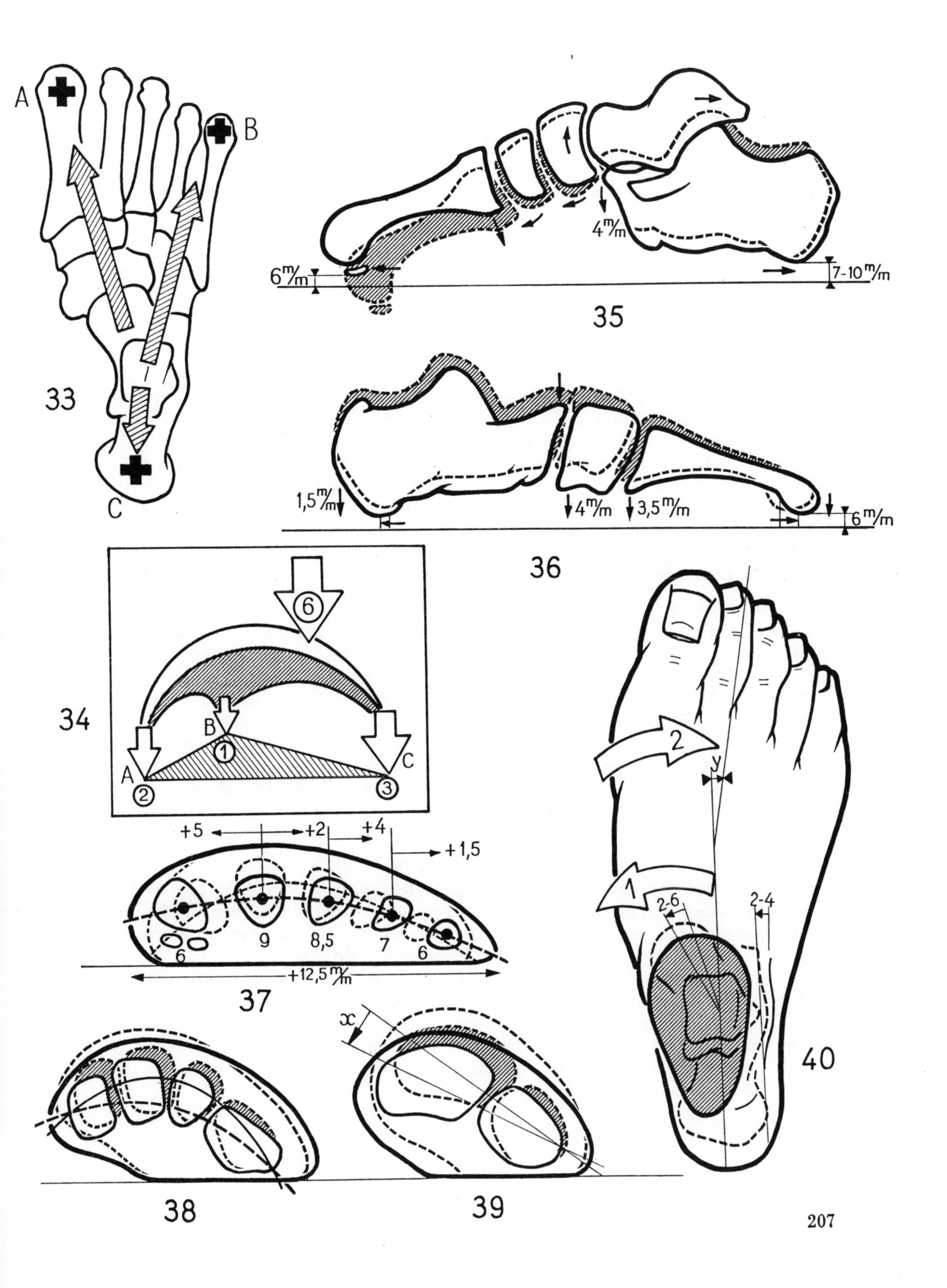
A
B
C
33
6 m/m
4 m/m
7-10 m/m
35
1,5 m/m
4 m/m
3,5 m/m
6 m/m
36
34
6
B
1
A
2
C
3
+5
+2
+4
+1,5
9
8,5
7
6
6
+12,5 m/m
37
x
y
2
1
2-6
2-4
40
38
39

DYNAMIC CHANGES OF THE ARCHES OF THE FOOT DURING WALKING

During walking the **evolution of the step** subjects these arches to stresses and strains, which highlight the function of these arches as an *elastic shock-absorber*. The 'unwinding' of the step has **four phases.**

Phase I: Contact with the ground is established (fig. 41).

When the forward limb (Vol. IV) is about to 'land' the ankle is straight or slightly flexed (fig. 41) by the action of the ankle flexors (F). The foot touches the ground **at the heel**, i.e. at the posterior support of the plantar vault. Straight away, under the thrust of the leg (white arrow), the foot is flattened on the ground (arrow 1) while the ankle is passively extended.

Phase II: Maximum contact (fig. 42).

The sole of the foot rests on the ground over the whole of its bearing surface which constitutes *the footprint*. **The body, propelled by the other foot, passes first vertically over the supporting limb and then moves in front of it** (period of unilateral support). Thus the ankle changes passively from the position of extension to a new position of flexion (arrow 2). At the same time, the weight of the body (white arrow) is fully applied to the plantar vault which is flattened. This flattening of the vault is simultaneously checked by contraction of the plantar tighteners (P) —the first stage of shock absorption. As it flattens, the vault is lengthened a little: at the start of this movement the anterior support (A) moves anteriorly slightly but at the end, when the anterior support becomes more and more firmly fixed to the ground under the weight of the body, the posterior support C, i.e. the heel, recedes. The surface area of the footprint is maximal when the leg is vertically above the foot.

Phase III: First stage of active propulsion (fig. 43).

The weight of the body is shifted to the anterior part of the supporting foot and **contraction of the ankle extensors** (T), especially **the triceps surae, raises the heel** (arrow 3). While the ankle is thus actively extended the plantar vault as a whole rotates about its anterior support (A). The body is lifted and moves anteriorly: this is the first stage of propulsion and is especially important as it depends on powerful muscles. Meanwhile, the plantar vault, caught between the ground anteriorly, the muscular force posteriorly and the weight of the body centrally (lever of the second type), would be flattened if the plantar tighteners (P) did not intervene: this is the second stage of shock absorption which allows some of the force of the triceps to be stored for release at the end of the propulsive movement. On the other hand, it is at the moment when the body is supported on the anterior part of the foot that the anterior arch is flattened in its turn (fig. 44) and the anterior part of the foot is splayed out on the ground (fig. 45).

Phase IV: Second stage of active propulsion (fig. 46 A and B).

The force provided by the triceps surae (fig. 46A) is followed by a second propulsive force (arrow 4) supplied by contraction of the flexors of the toes (f), especially the flexor hallucis brevis, the adductor and the abductor hallucis and the flexor hallucis longus. The foot is now once more raised further anteriorly and is no longer supported by the anterior tarsal bones; it now rests entirely on the first three toes (fig. 46B), especially the big toe—the final stage of support (A′). During this second propulsive movement the plantar vault resists flattening once more thanks to the plantar tighteners, comprising the flexors of the toes *inter alia*. It is at this stage that the energy stored by these tighteners is released. The foot then leaves the ground while the other starts a new step. Thus both feet are in contact with the ground for a very short time (period of bilateral support). In the next stage—i.e. of unilateral support—the vault of the foot which has just left the ground moves back to its original position.

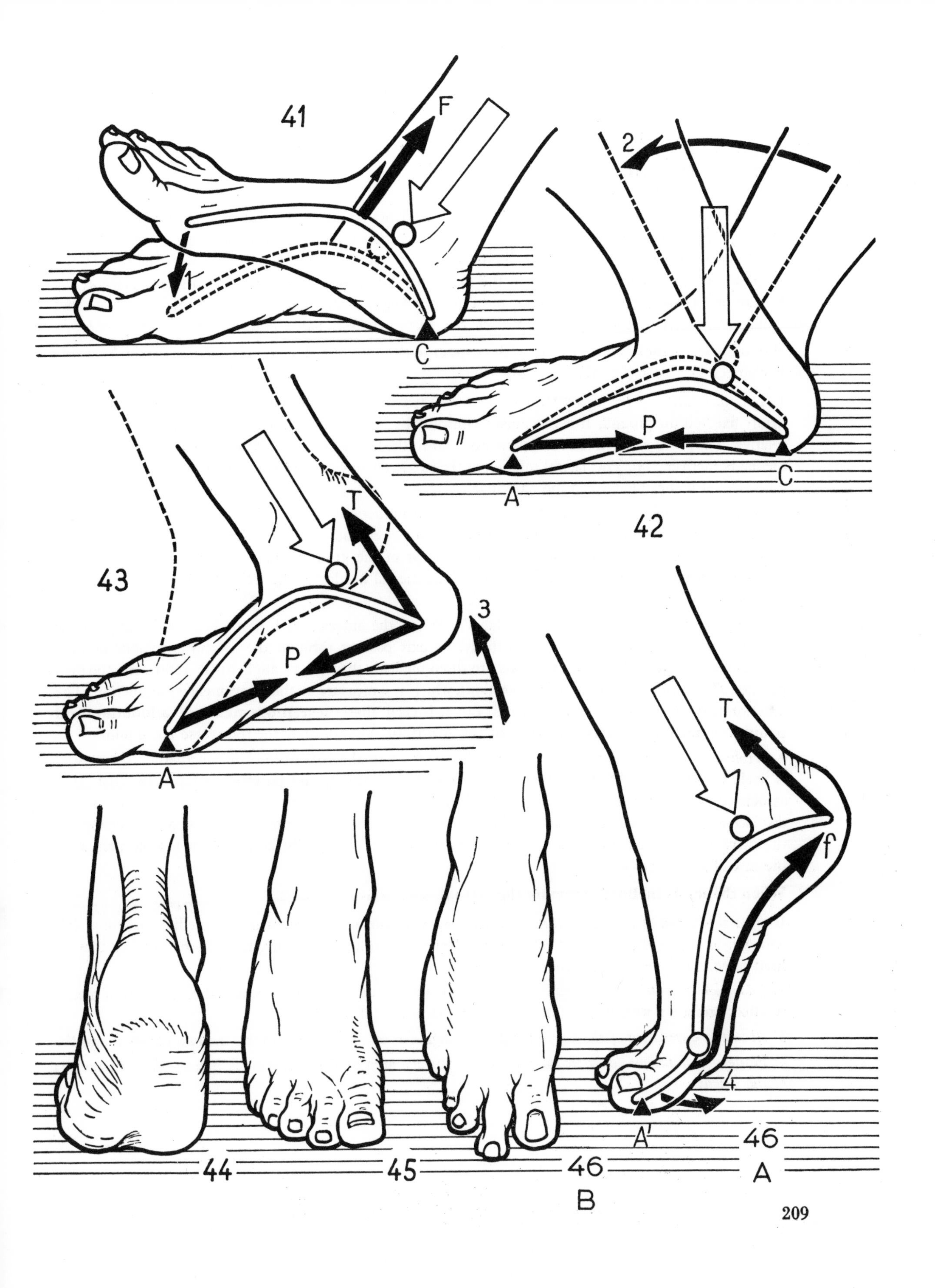

41
F
1
C
2
P
A
C
42
43
T
P
3
A
T
f
4
A'
46
A
44
45
46
B

DYNAMIC CHANGES RELATED TO THE MEDIAL AND LATERAL INCLINATION OF THE LEG ON THE FOOT

So far we have studied the alterations occurring in the plantar vault during walking, i.e. following changes of the angle between leg and foot in *the sagittal plane*. During walking or running along a *curved track or uneven ground* it is essential that the leg should be able to change the angle it forms with the foot in **the frontal plane**, i.e. medial and lateral to the footprint. These side-to-side movements take place at the **subtalar and transverse tarsal joints** and lead to changes in the shape of the plantar vault. Note that the ankle is not involved: *the talus, gripped between the two malleoli, moves relative to the other tarsal bones.*

When the leg is inclined medially with respect to the foot (considered to be stationary) the following four changes take place (fig. 47):

1. *Lateral rotation of the leg on the foot* (arrow 1) which takes place only when the sole of the foot is firmly fixed on the ground. It is recognised clinically by the posterior movement of the lateral malleolus relative to its position when the foot, perpendicular to the leg, is in contact with the ground only along its medial border (fig. 48). This lateral rotation of the two malleoli leads to a lateral displacement of the talus, especially of its head lodged in the navicular.

2. *Abduction-supination of the hindfoot* (fig. 49). The abduction is due to an uncompensated component of lateral rotation. Supination results from medial displacement of the calcaneus which is obvious from the back (angle x) and when referred to a foot clear of the ground (fig. 50). This 'varus' movement of the calcaneus is associated with a change of the medial border of the Achilles tendon from straight to concave.

3. *Adduction-pronation of the forefoot* (fig. 47). When the anterior arch is applied to the ground the anterior part of the foot is displaced medially: its axis passing through the second metatarsal and the sagittal plane P passing through this axis are tilted medially through an angle m (P′ represents the final position of this plane and P its initial position), which is a measure of this movement of adduction. The anterior part of the foot is also pronated. But it is clear that **these movements of adduction and pronation are only relative to those of the hindfoot.** They occur at the transverse tarsal joint.

4. *'Hollowing' of the medial arch* (fig. 47). This increase in the curvature of the medial arch (arrow 2) is itself a consequence of the relative movements of the anterior and posterior parts of the foot. It is associated with elevation of the navicular relative to the ground: this elevation is at once passive (lateral displacement of the talar head) and active (contraction of the tibialis posterior). The overall change in the curvature of the plantar vault is seen in the change in its outline: the hollow of the foot deepens as in the case of a *pes cavus varus.*

When the leg is inclined laterally (fig. 51) the exact converse occurs:

1. *Medial rotation of the leg on the foot* (arrow 3): posterior displacement of the medial malleolus (cf. with fig. 52 where the foot only rests on its lateral margin); medial movement of the talus so that its head projects on the medial border of the foot.

2. *Adduction-pronation of the hindfoot* (fig. 53): adduction due to an uncompensated component of medial rotation; pronation with valgus (angle y) of the calcaneus (cf. fig. 54).

3. *Abduction-supination of the forefoot* (fig. 51): angle (n) of abduction between the two planes P and P′.

4. *Flattening of the medial arch* (arrow 4): the surface area of the footprint is increased, as in the *pes planus valgus.*

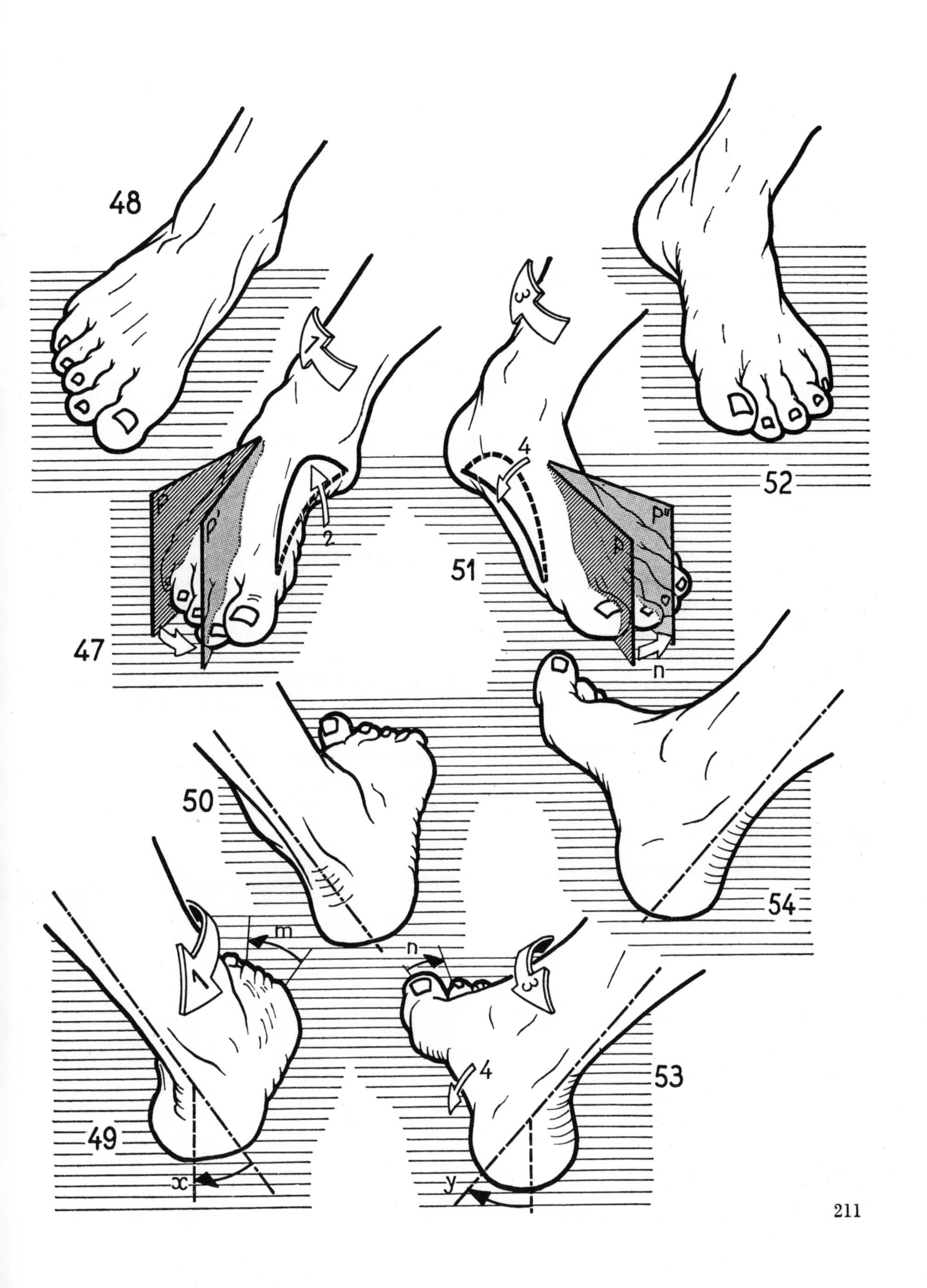
48
1
2
P
P'
47
3
4
P
P"
n
52
51
50
m
1
49
x
54
n
3
4
53
y

ADAPTATION OF THE PLANTAR VAULT TO THE GROUND

The town dweller always walks on even and firm ground with his feet protected by shoes. There is therefore little need for the arches of his feet to adapt to new terrains and the supporting muscles eventually atrophy: the flat foot is the price paid for progress and some anthropologists go so far as to forecast that man's feet will be reduced to mere stumps. This thesis is borne out by the fact that in man in contrast to the ape the toes are atrophied and the big toe can no longer be opposed.

This stage is still to come and even 'civilised' man can still walk barefoot on a beach or on the rocks. This return to the primitive state is highly beneficial to the plantar vault (*inter alia*), which thus retrieves its adaptive capabilities.

Adaptation to the uneven features of the ground, which the foot can grasp within the hollow of the vault (fig. 55).

Adaptation to slopes of the ground with respect to the feet: the bearing surface of the forefoot is more extensive *when the ground slopes laterally*, because of the decreasing lengths of the metatarsal bones mediolaterally (fig. 56);

when one *stands on a transverse slope* (fig. 57) the foot 'downstream' is in supination while the foot 'upstream' is everted or in talus valgus;

in climbing (fig. 58), the foot 'downstream' must be firmly anchored to the ground perpendicular to the slope, i.e. in a position of pes cavus varus while the foot 'upstream' approaches the ground in maximal flexion and parallel to the slope;

in coming down a slope (fig. 59) the feet must often be inverted so as to secure maximum grip.

Thus, just as the palm of the hand allows prehension by changing its curvature and its orientation in space (Vol. I), the sole of the foot can within limits adapt to the irregularity of the ground so as to ensure optimal contact with it.

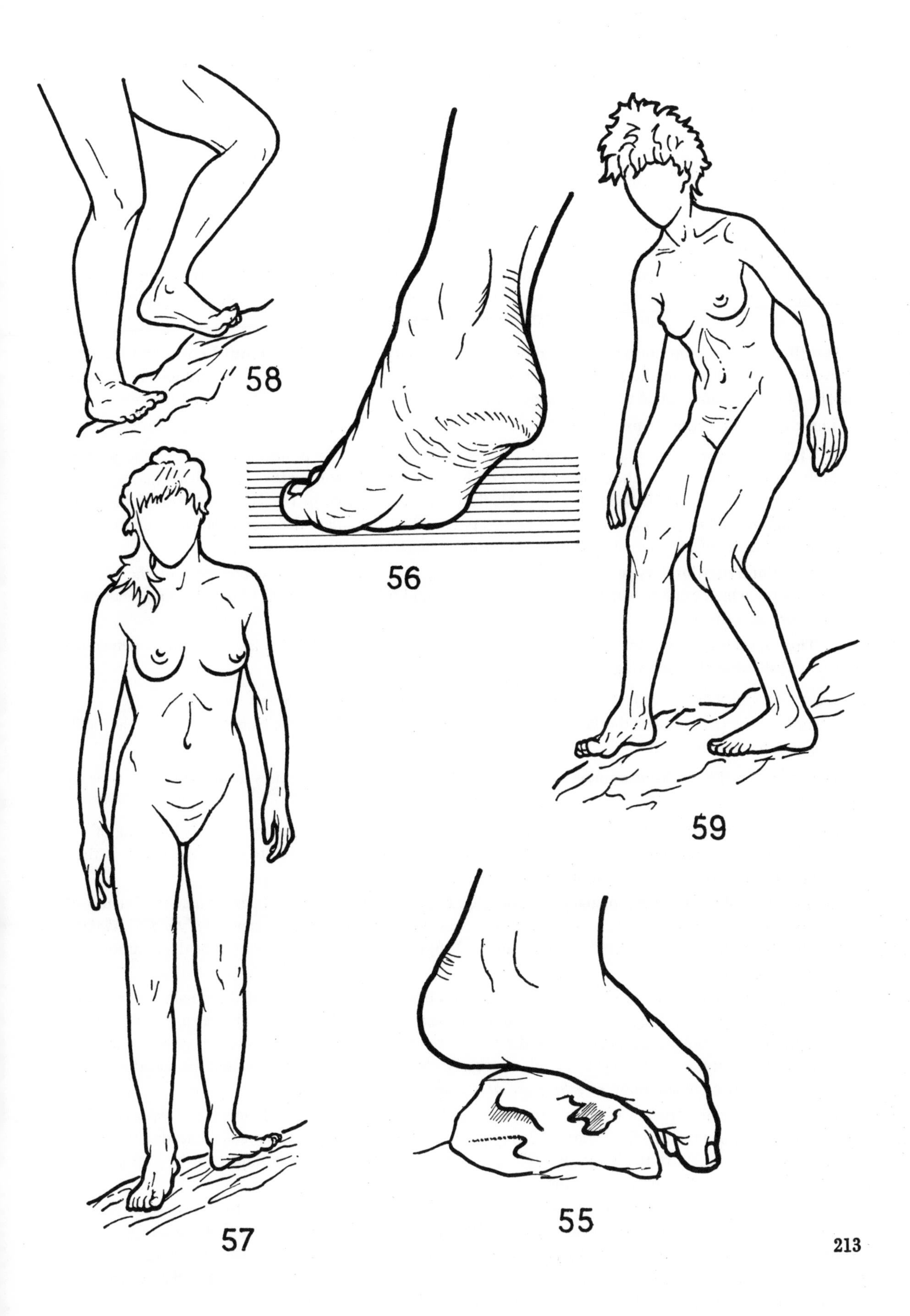
58
56
59
57
55

CLAW FEET (PES CAVUS)

The curvature and orientation of the plantar vault depend upon a very delicate balance of the various muscles concerned. This can be studied with the help of Ombrédanne's model (fig. 60):

the vault is flattened by the weight of the body (white arrow) and by contracture of the muscles attached to the convexity of the vault: the triceps surae (1), the tibialis anterior and the peroneus tertius (2), the extensor digitorum longus and the extensor hallucis longus (3). The last two muscles are effective only if the proximal phalanges are stabilised by the interossei (7).

the vault is 'hollowed' by contracture of the muscles attached to its concave aspect: the tibialis posterior (4), the peroneus longus and brevis (5), the plantar muscles (6) and the flexor digitorum longus (8). It can also be hollowed by a relaxation of the muscles inserted into its convexity. Conversely, relaxation of the muscles in its concavity leads to a flattening of the vault.

Insufficiency or contracture of a single muscle disrupts the overall equilibrium and leads to some deformity. Duchenne de Boulogne states in this connection that it is better to have all the muscles paralysed rather than a single muscle, since then the foot retains a fairly normal shape and position.

There are **three types of pes cavus**:

1. **The 'posterior' type** (fig. 61), where the deformity involves the posterior buttress of the plantar vault due to insufficiency of the triceps surae (1). The muscles in the concavity of the arch are inadequately balanced (6) and the sole is hollowed; the ankle flexors (2) tilt the foot in a position of flexion. This leads to the *talipes equinus* (fig. 62) which is often compounded with a *valgus* deformity following contracture of the abductors of the foot (the extensor digitorum longus, the peronei muscles).

2. **The 'intermediate' type** (fig. 64) is relatively rare and results from contracture of the plantar muscles (6), which can follow the use of shoes with too rigid soles or shortening of the plantar aponeurosis (Ledderhose's disease).

3. **The 'anterior' type** can be further divided into subgroups which all share an equinus deformity (fig. 65) showing the following two characteristics:

equinus deformity of the forefoot (e) due to lowering of the anterior buttresses of the vault,

a *misalignment* (d) between the heel and the forefoot, which can be partially reduced when the body weight is being supported.

Depending on the mechanism underlying the condition, the following varieties of **the anterior type of pes cavus** are described:

contracture of the tibialis posterior (4) and of the peronei longus and brevis (5) causes a lowering of the anterior part of the foot (fig. 66). Contracture of the peronei alone can lead to pes cavus which is then compounded with a valgus deformity, i.e. *talipes arcuatus*, *equinovalgus*;

an imbalance of the metatarsophalangeal joints (fig. 68) is a common cause of pes cavus: insufficiency of the interossei (7) tips the balance in favour of the toe extensors (3) and *hyperextension of the first phalanx* follows. Next the metatarsal heads become lowered (6) and this leads to a lowering of the anterior part of the foot; hence pes cavus;

lowering of the metatarsal heads can also be due (fig. 69) to an insufficiency of the tibialis anterior (2): the extensor digitorum longus (3) attempts to compensate and tilts the proximal phalanges; the plantar muscles (6), now unbalanced, accentuate the curvature of the vault and triceps action provokes a slight equinus deformity. A slight degree of valgus (fig. 70) follows from the inadequately balanced extensor digitorum; hence the condition of *talipes arcuatus equino-valgus*;

a common cause of clawfoot is the wearing of shoes that are too short, or of *high-heeled shoes* (fig. 71): the toes hit against the tip of the shoes and are hyperextended (a) so that the metatarsal heads are lowered (b). Under the weight of the body (fig. 72) the foot slides forward down the slope and the heel and toes are approximated. This exaggerates the curvature of the vault.

The diagnosis of claw foot is made easier by studying the **footprint** (fig. 73). In comparison with the normal footprint (I) the first stage of pes cavus (II) shows a projection on its lateral border (m) and a deepening of the concavity of the medial border (n). The next stage (III) shows a footprint which is divided into two. Finally, in the long-standing cases (IV), these characteristics become associated with the disappearance of the prints of the toes (q) due to a secondary claw-toe deformity.

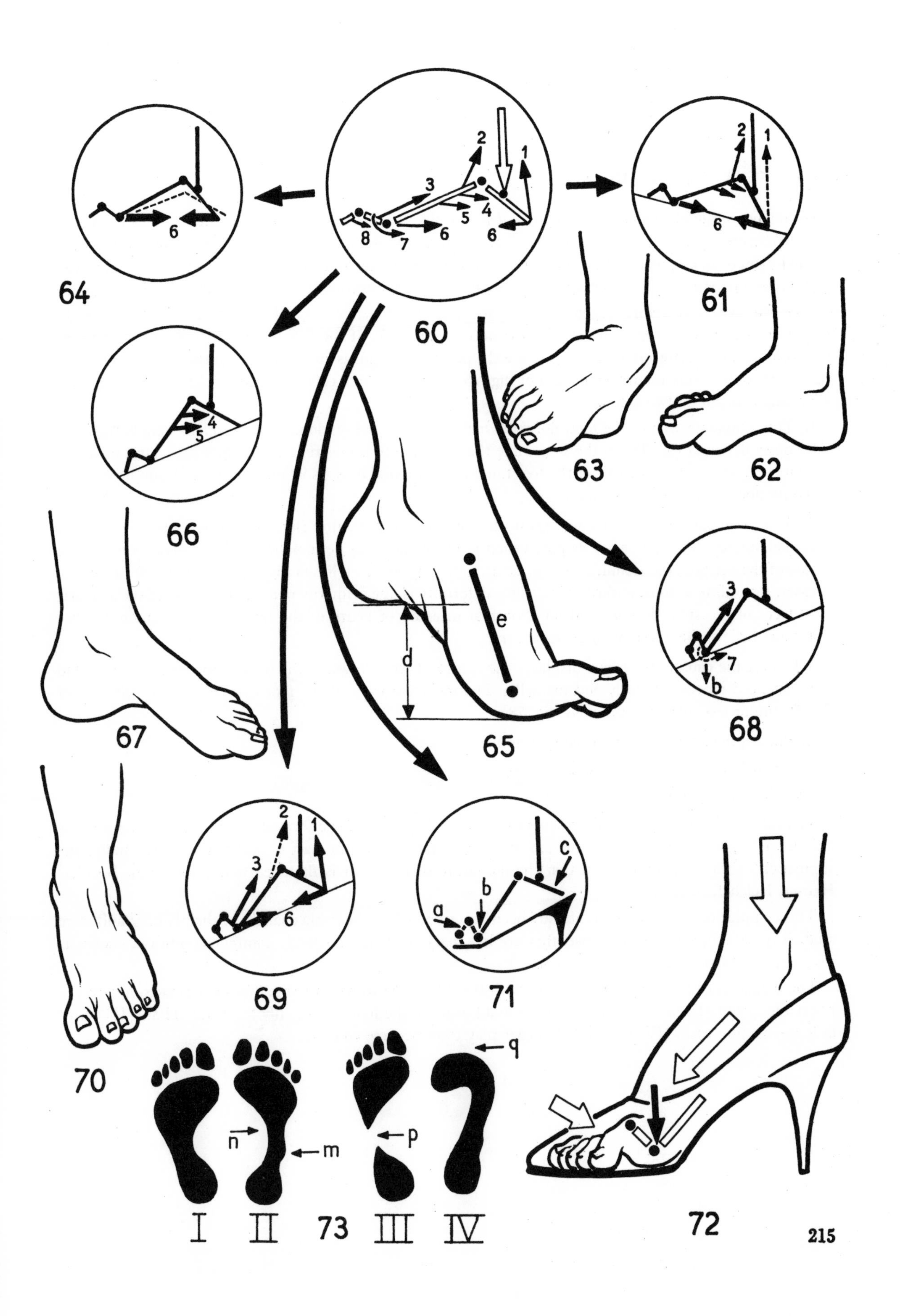

1
2
3
4
5
6
7
8
64
60
61
66
63
62
67
e
d
65
68
b
69
71
a
c
70
q
n
m
p
I
II
73
III
IV
72

FLAT FEET (PES PLANUS)

The collapse of the plantar vault is due *to weakness of its natural means of support*, i.e. muscles and ligaments. The ligaments by themselves are capable of maintaining the integrity of the vault for a short period since the footprint of an amputated leg is normal except if the ligaments have been previously cut. In life however, if the muscular support fails, the ligaments become stretched eventually and the vault collapses for good.

The flat foot is therefore due mainly to muscular insufficiency (fig. 74): insufficiency of the tibialis posterior (4) or, more commonly of the **peroneus longus** (5). If the foot is not supporting the body the foot shows a varus deformity (fig. 75) because the peroneus longus is an abductor. On the other hand, when the weight of the body is applied to the foot (fig. 76) the medial arch collapses and a *valgus deformity results.* This valgus is due to two factors:

1. The transverse arch of the foot, normally maintained by the tendon of the peroneus longus (fig. 77), becomes flattened (fig. 78); at the same time the medial arch is lowered: the forefoot (e) rotates medially on its long axis so that the sole of the foot touches the ground over its whole surface and simultaneously the forefoot is displaced (d) laterally.

2. The calcaneus turns on its long axis in the direction of pronation (fig. 79) and tends to lie flat on its medial surface. This degree of valgus, which is visible and can be measured by the angle between the axis of the heel and the Achilles tendon, exceeds the physiological limits (5°) and can attain 20° in certain cases. According to some authors, this valgus deformity is due primarily to a malformation of the articular surfaces of the subtalar joint and an abnormal measure of laxity of the interosseous ligaments; other authors believe these lesions to be secondary.

Whatever the cause, this valgus displaces the centre of stress towards the medial border of the foot and the talar head moves inferiorly and medially. The medial margin of the foot then shows the presence of three more or less distinct *projections* (fig. 78):

the medial malleolus (a), abnormally prominent;

the medial part of the head of the talus (b);

the tubercle of the navicular bone (c).

The tubercle of the navicular represents the apex of the obtuse angle formed by the axes of the posterior and anterior parts of the foot: adduction-pronation of the posterior part is compensated by abduction-supination of the anterior part, so that the curvature of the vault is flattened out. (Hohmann, Boehler, Hauser, Delchef, Soeur).

This complex of deformities has already been described when the static changes of the plantar vault were studied (p. 207, fig. 40). (In this case, they are less marked.) It is a relatively common condition, known as *the painful flat foot or tarsalgia of the young.*

The diagnosis of flat foot is made easier with the use of the **footprint** (fig. 80): in comparison with the normal footprint (I), the concavity of the medial border of the foot is gradually filled out (II and III) until in long-standing cases (IV) the medial border may even become convex.

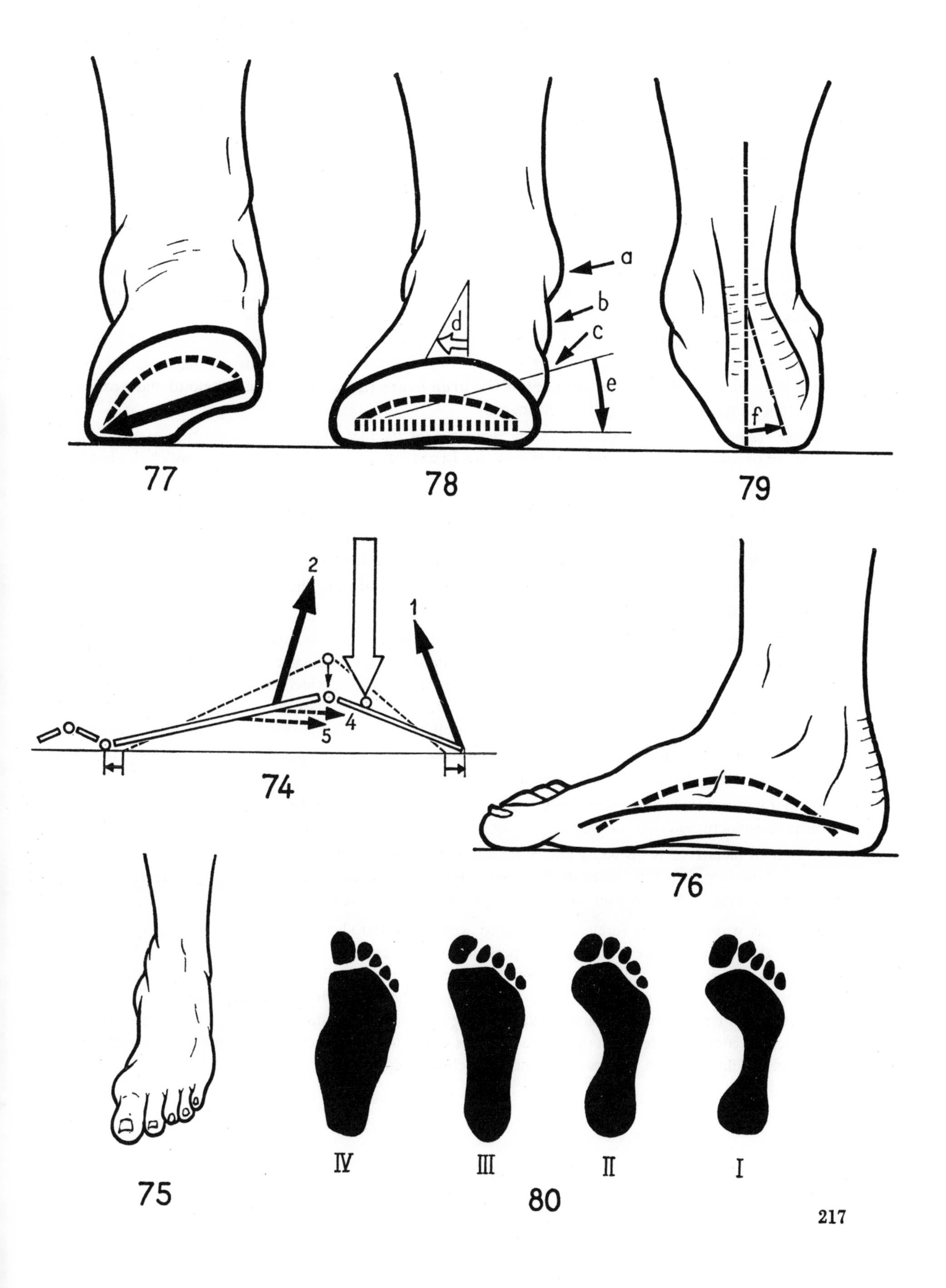
a
b
c
d
e
f
77
78
79
2
1
4
5
74
76
75
IV
III
II
I
80

IMBALANCE OF THE ANTERIOR ARCH

In the development of deformities of the plantar vault the balance of the anterior arch can be upset at the level of its **supports** and by changes of its **curvature**.

This imbalance is generally secondary to the anterior type of pes cavus: the equinus deformity of the forefoot enhances the stresses applied to the anterior arch in the following three ways:

1. *The equinus deformity of the anterior part of the foot is symmetrical* (fig. 81), i.e. without any pronation or supination of the foot; the curvature of the arch is preserved. Thus the **two supports become overloaded** and callosities develop under the heads of the first and fifth metatarsals.
2. *The equinus deformity is associated with pronation of the foot* (fig. 82) due to a greater degree of lowering of the medial arch (contracture of the tibialis posterior or of the peroneus longus); as the curvature of the arch is maintained, **the medial support of the arch bears the brunt of the overload** and a callosity develops under the head of the first metatarsal.
3. *The equinus deformity is accompanied by supination of the foot* (fig. 83): the curvature of the arch is maintained and the **lateral support bears the brunt of the overload** (callosity under the head of the fifth metatarsal).

In certain types of anterior pes cavus the curvature of the anterior arch can be altered as follows:

partially flattened or *straightened* (fig. 84): this is the case of the **anterior type of flat foot**; the overload is distributed to all the metatarsal heads (callosity under each head);

completely reversed (fig. 85): this is the **anterior type of round foot**; the overload is borne by the heads of the three middle metatarsals (with corresponding callosities).

Inversion of the anterior arch is due to a claw or hammer deformity of the toes. As we have seen before, this deformity of the toes can result from an imbalance between the interossei and the extensors; very often it results from shoes that are too short or from high heels (which are equivalent to tight shoes): the toes (fig. 86) hit against the front of the shoes and are bent; the head of the first phalanx is pulled down and a callosity appears; the head of the metatarsal is itself lowered (callosity develops) and the arch collapses.

This also occurs readily when pointed shoes are worn on feet of a particular configuration: the **pes anticus** (the Neanderthal foot) which recalls the prehuman foot with the prehensile big toe (fig. 87):

the first metatarsal is short, excessively mobile and set far apart from the second metatarsal (metatarsus varus or adductus) so that the big toe runs obliquely anteriorly and medially;

the second metatarsal is distinctly longer than the others so that it supports the weight during the final phase of the step. It is thus overloaded and pain develops at its base; occasionally fatigue fracture occurs; the fifth metatarsal is widely set laterally (valgus deformity of the fifth metatarsal).

When this widely splayed forefoot is confined within pointed shoes (fig. 88) the big toe is displaced laterally (a). This imbalance soon becomes permanent as a result of shortening of the capsular ligament of the joints, lateral dislocation of the sesamoid bones (c) and of the tendon, the formation of an exostosis (b) and of a callosity on the metatarsal head. This is the pathogenesis of the **hallux valgus**. The big toe displaces the intermediate metatarsals exaggerating their hammer deformity (fig. 89). The fifth toe undergoes the converse deformity: this is the **quintus varus** which further enhances the hammer deformity of the intermediate toes. In this way the anterior arch becomes convex.

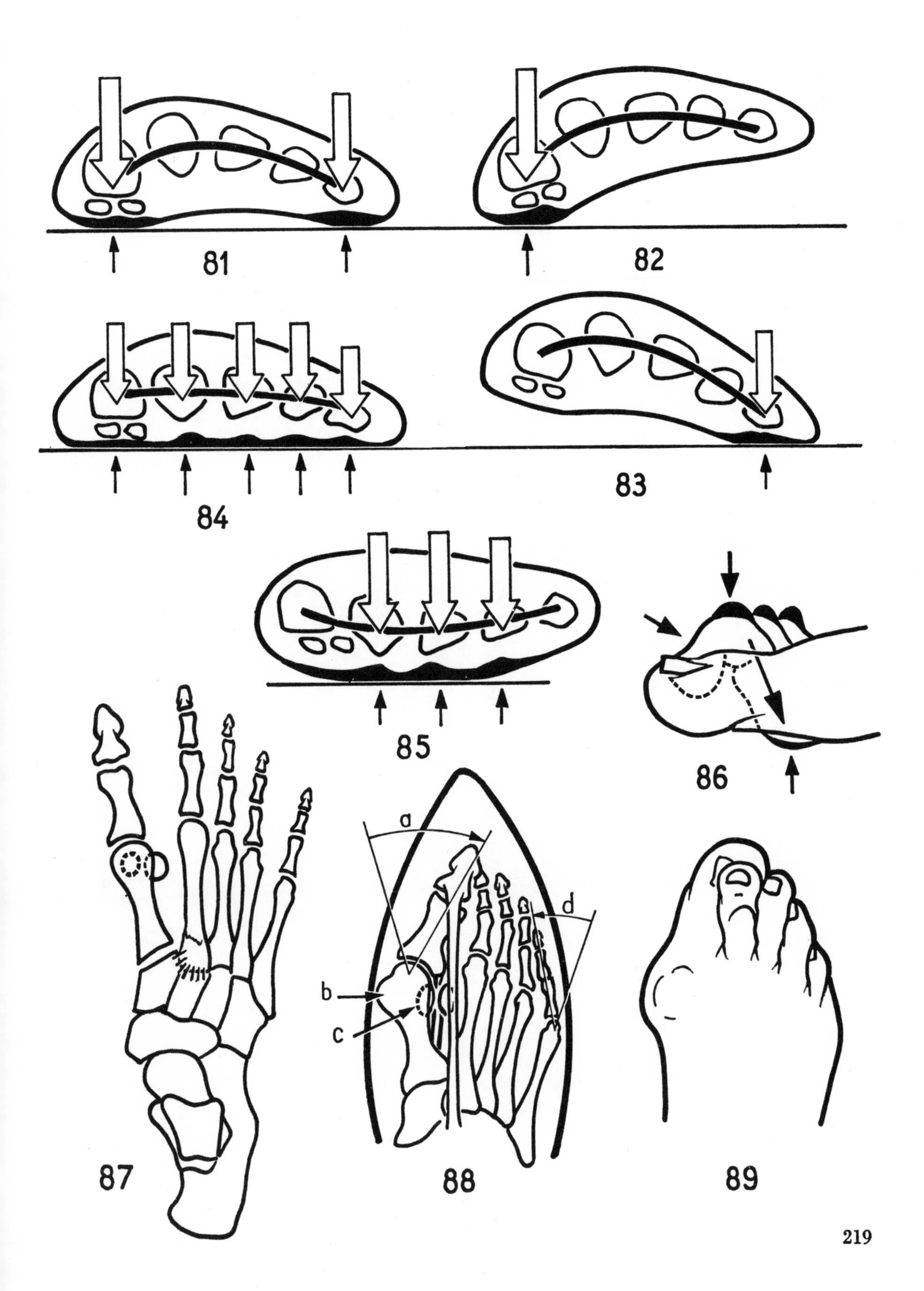
81
82
84
83
85
86
87
88
89
a
b
c
d